MW01248619

FROM SHADOWS and SYMBOLS to the TRUTH
Cor ad Cor loquitur
MARYGROVE COLLEGE
EX LIBRIS

LIFE AND LETTERS OF WALTER DRUM, S.J.

WALTER DRUM, S.J.

THE LIFE AND LETTERS

OF

WALTER DRUM, S.J.

BY

JOSEPH GORAYEB, S.J.

PREFACE BY

FRANCIS P. LEBUFFE, S.J.

THE AMERICA PRESS

NEW YORK

1928

PREFACE

It is not often that the writer of a biography follows the subject of his writing within the shadows of death before he has seen his work in print. Yet this has been the fate of the author of this life, Father Gorayeb, and his work, spelled out in devotion to his friend, must now be edited by one who was privileged to call himself a friend of the writer and a still closer friend of the subject whose life-story is herein told. Both were sons of Ignatius and both fought well beneath his banner, and now rest with their soldier-brethren, one at St. Andrew-on-the-Hudson, the other at Woodstock.

Father Gorayeb had all but completed the long course of Jesuit training and had just finished his "long retreat" of thirty days when he was stricken with his final sickness. He went home to God before he had time to wage battle in active warfare for Our Lord. But Father Drum, of whom Father Gorayeb writes so well, had for years drawn the fire of the enemy in his front-line warfare and had spent himself in relentless campaigning for his Captain, Christ. The young soldier and the older warrior are now bivouacked with the unnumbered hosts who have "preceded us with the sign of Faith and sleep the sleep of peace." May God rest their warrior souls!

Of the subject of this biography little need be said, for Father Gorayeb has written all and written it tellingly. Father Drum's character was a dominating one and stood out imperatively in every gathering of men. He had his faults, as every child of Adam has, but "a diamond with a flaw in it is better than a common stone with none." By temperament—that gift of nature which is ours by birth—he was imperious and proud; but by character, and thus in deliberate action, he was humble and lowly-minded. At times, when taken unawares, nature would assert itself and the latent pride would show itself, but only to provoke meek apologies when the outbreak was detected. This "dual personality," proud by temperament and humble by character-formation, was the explanation of the varying judgments passed on a man about whom every one who met him was forced to form an opinion. His temperament showed itself in the pounding military walk, in a marked *hauteur* which was at times stamped out largely on his actions, in certain mannerisms that bespoke self-consciousness, if not self-esteem; his character was seen as he knelt in prayer, (echoes of which we catch in his diary and his letters), in his readiness to perform any task assigned, in his love of poverty, in his zeal for souls and his spendthrift activity that he might bring them nearer to his Captain Christ, whom he loved with all the tremendous passion of the Ignatian "banner-man." He literally spent himself for Christ. He must be up and doing always, and of him St. Paul's words were splendidly true—"the love of Christ drives us on." It did drive him on until "taps" sounded and Christ's soldier stacked his arms, and the fever of life was over—and rest came.

FRANCIS P. LEBUFFE, S.J.

Campion House,
Feast of Nativity, B.V.M., 1928.

Table of Contents

INTRODUCTION

All the world loves a soldier. Though we have had a surfeit of war, and of war's aftermath, yet there will ever remain in us an undercurrent of admiration—for the erect bearing, the firm, sure step, the alertness and precision and efficiency that always catch the eye in a soldier. We conjecture, too, that uprightness and strength and fearlessness are of the character itself, that these fine qualities enter into every detail of the soldier's life, his occupation and his training.

The picture is still more engaging in the higher service of Christ. The same qualities are heightened, and hallowed, till they acquire in our eyes the attributes that we not only admire but reverence in a priest of God.

This book is a modest attempt to portray such a character: a soldier's son, a soldier's brother, himself a soldier in a yet higher service. But it is a difficult task to describe every facet in the many-sided character and personality of such a man as Father Drum, and to conjecture what he might have been, what greater good he might have accomplished had God seen fit to lengthen his career. However, although he lived enough to accomplish a deal of good for his living, it is not so much the fact of actual accomplishment that has made it seem worth the while to study the man and his work. It is rather the sum total, the ensemble of a life perhaps too rare in these times, that of a man who was, if anything, tremendously earnest in his loyalty; loyalty to the cause of Christ, loyalty to his friends, loyalty to the Order of which he was a member, with a loyalty that manifested itself to the least and last detail. He was a soldier, whether as a young lad eager to follow in the footsteps of his soldier father; or, when the higher call came, as a youthful recruit giving himself wholeheartedly, with military thoroughness and precision, to his every duty in Religious life; or later, as a scholarly priest, ever endeavoring to carry out the same efficient principles of soldierly service, with a zeal and fervor and absorption that only absolute loyalty could have sustained through long laborious days.

It is this example of the soldier in him, that has made

possible the hope that, in presenting Father Drum as he was, his life and character and work may prove to be a help and an inspiration to others who are battling in the same grand cause, or who wish to offer themselves *ad insigniora*. At least, the labor will have found ample reward enough, if this little sketch may be a help to his many friends to treasure with profit a holy and Christ-like memory.

"Blessed is the Byzantine nature," Father Drum thus describes the type of character which attracted him most, in terms of what he came to regard as his favorite style of architecture:

> How impressive is the great dome of Aia Sophia, St. Sophia, of Constantinople! It is set upon half domes, which are in turn supported by four other half domes. This great dome, upon domes supporting domes, upheld by gigantic adamantine piles, gives one an intense realization of the power, the massive grandeur, the overwhelming and overshadowing magnificance of the Church of Christ! Such are some characters—great, grand, full, massive, generous, inspiring. What a pleasure it is to meet such a one! Grace can readily enough build on the Byzantine nature. And the ennobling of the noble qualities of the Byzantine character results in a most inspiring and encouraging large-mindedness and great-heartedness in the service of God.

Acknowledgment is due, for much generous assistance in the work of preparing this sketch, to a Sister of Notre Dame de Namur, Worcester, Mass.; to a Visitandine, Baltimore, Md.; to Miss Ethel Cuttle, Fall River, Mass.; to Miss Jeannette T. Brotherton, Catonsville, Md.; and to many other friends for valuable services; and to Mr. John D. Drum, brother of the subject of this biography, for courteous cooperation in the collation of material and the verification of data. Only lack of space has necessitated the omission of much other material that was generously left at the editor's disposal.

CHAPTER I

THE BATTLE CRY

It is a commonplace of biography to find that somewhere along the line of years, at some point in the career of every man, an event occurs which seems to define his whole life, and shape his character and his destiny. At the time, the fact may not stand out so clearly. The perspective of later years must throw the event into its true alignment. It may be crystallized into a single utterance, that for him becomes thenceforth the guiding principle of life. Some such significance appears to have been given to that brief half-hour that Father Drum was once privileged to spend, in private audience with His Holiness Pius X. The audience took place at 11.30 a. m. on September 30, 1907, during a brief stay in Rome while he was en route for his course of studies at Innsbruck. His own words will best describe this experience:

> There was no difficulty at all in getting a private audience. Mgr. Bisleti, the Majordomo, is most cordial. His Secretary took the letter of introduction I had brought from Father Mullan, and returned at once to usher me into the Monsignor's study. Mgr. Bisleti spoke highly of the work of the Society in Biblical Studies, especially of the work of Father Fonck, whom the Monsignor hopes soon to see on the Biblical Commission. When I explained that I wished an audience, I was told the Holy Father would take great pleasure in talking to me about my studies. Two or three days later the invitation was brought me by special messenger.
>
> Father Carroll of the California Mission, was with me during my audience. His name was not on the invitation, and we feared some difficulty. There was none. Whenever we were held up by any of the guards or attendants, we routed them by a business-like wave of the hand, and the magic word, "udienza," and forged onward. When we reached the waiting-room we were dismayed by the hard-heartedness of the ushers in rose-colored knickerbockers; they said it would be impossible to usher in anybody but me, as the invitation called for no one else. However, at our request, they called a Monsignore; he took the invitation away, and very shortly

returned with the kind information that the Holy Father would receive us, either both together or one at a time. When our turn came we entered together at 11.30 a. m.

That entrance I shall never forget. I had expected to find myself in the throne-room; and had been told that a Monsignore would put me through the motions of the ceremonial. There was nothing like that. We were in the Pope's study. He stood as we entered. I looked for the Monsignore; he was gone. So down I went on both knees and made the solemn bow. At once, a firm, kind voice said: *Surgite, accedite, sedete.* The Holy Father is above ceremonial; his simplicity and strength and deep knowledge of human nature often led him to lay aside the time-honored formalities of the Vatican. Still, I felt the three genuflections were his due, and was determined to give them. So I made a second and simple genuflection. The Holy Father said: *Sedete, sedete.* I then knelt at the feet of His Holiness and received his blessing. He would not allow me to kiss his feet, but made me sit in a chair close by his desk. We chatted for nearly half an hour. As the audience had been arranged for me, my companion left me the floor; I took it fearlessly.

That talk of half an hour with Pius X is one of the greatest experiences of my life. The memory of that dear old man; of his piercing, telling, yet kindly eyes; of his countenance, at times bright and humorous, at times sad and solicitous, but always strong; of his words, authoritative, inspiring, heartfelt and heart-reaching, that memory will ever be an occasion of grace, such as no other experience but Ordination and First Mass will be.

During my entire audience with the Holy Father, it was clear to me, how he esteemed the Society and how he counted on it. On one occasion he said slowly and with heart: *La Compagnia é bene merita*; then he gazed at me with sad eyes and sorrowful, as he said: *bene merita della Chiesa di Dio.*

Most of our talk was about my studies. With all a father's interest, sincerity and pride, he asked me in detail about my studies in the Society. Had I made the four years of theology? the three years of philosophy? the tertianship? He was much pleased that I had the full training of the Society before starting to specialize in Biblical Studies. "Ah, 'tis well," said he, "you are a full-formed Jesuit."

Then we talked about my studies and travels in Syria and Palestine. The dear old man wished to know what

Semitic Languages I had studied, how long I had studied each, and so forth. He inquired much about our Oriental Faculty at Beirut. When I told him that last year there were twenty-five Jesuits studying Arabic in Beirut, his face glowed with joy. "Those Jesuits," he said "what won't they do!"

When I spoke of my coming studies at Innsbruck, the Holy Father referred with pleasure to Fr. Fonck as to a man who used modern science to uphold traditional truth.

In regard to Biblical work, the grand old man spoke ever firmly and unflinchingly. There was but one burden to his words: "Fight for the traditional teachings of the Church!" He was sad yet indignant, as he spoke of the Modernists. It was clear he meant war with them to the finish. The day of my audience was the feast of St. Jerome. After one of the Holy Father's outbursts against the Modernists—whom he always referred to as "those wiseacres"—I ventured the suggestion that the office and works of SS. Jerome and Augustine entered less into the lives and studies of the Modernists than did the writings of *Il Santo* Delitzch and *Il Santo* Harnack. The Holy Father chuckled at the allusion and struck his desk with a chubby fist as he said: "Yes, because they have read a page or two of some German rationalist, they think to overturn the teachings of all the Fathers of the Church, —*quelli sapientini!*" I thought of St. Paul's *solicitudo ecclesiarum!*

When the talk was over, I pulled out a large photograph of the Holy Father, which I had stowed away under my cassock. The Monsignoré take away such photographs if they see them. The Holy Father laughed as I tugged away at the smuggled treasure, and most willingly wrote upon the precious souvenir. He blessed everything I presented. Then I began to ask for personal blessings. First came blessings for the Province, Fr. Provincial, Woodstock, and the novitiate. The blessings were granted each with a hearty consent. I then asked a special blessing for my poor old mother. Pius X showed he is a man's man, a Pope with the feelings of a man. "Tell your dear old mother, I bless her with all my heart; tell her how I thank her for giving you to the Church, and to the Society of Jesus." My heart was full of enthusiasm, and my eyes were more than moist.

After he had blessed me, as I knelt before the dear old man, he got deep into my heart by one more human incident. "And now," said he, "let us make the intention by which each of us gives the other a share in all his future Masses." I bent down to show my humble thanks by kissing the feet of Our Lord's Vicar. He drew back a bit, and I, in ignorance of the authority of my words, insisted: *Voglio, Santo Padre.* The supreme jurisdiction of the Church yielded to the whim of a lowly follower; I was satisfied. One more blessing was received and I made for the door. As I looked back to say good-bye, the Holy Father hit his desk and urged: *"REMEMBER, STAND BY THE TRADITIONAL DOCTRINES OF THE CHURCH."* "Holy Father," I made answer, "if ever any argument of a rationalist seems to me strong, my memory of the talk with Your Holiness will by God's grace give me strength to hold my ground, and fight for the traditional teachings of the Church." The dear old man stood at his desk in kindly smiles, and we left the room.

The enthusiasm of that day I shall never forget. Pius X inspires one with the old crusader spirit.

CHAPTER II

MILITARY ANNALS

> "The life of Captain Drum is an inspiration to the men who knew him, living, and are now giving to him, dead, the last solemn tribute of honor and reverence; an inspiration to lead an upright, pious, honorable life, and thus be certain that their death, like his, shall be one of glory."

It was at the close of the Mass of Requiem, and the end of a moving tribute from the preacher of the occasion, the Rev. John F. Quirk, S.J., at a military funeral in the Church of the Immaculate Conception, Boston; at a time when the press of the country was blazoning the stirring news of the war with Spain, and all America was flushed with the excitement of victories in Cuba and on the sea. Now and then, however, men paused to weigh the solemn import of glory's aftermath. The morning of Saturday, September 3, 1898, was one such occasion. The body of the gallant Captain John Drum, which had been brought from its temporary resting place at the foot of San Juan Hill, a short distance beyond which he had fallen in battle, on the preceding first of July, and which had lain in state, for a brief day, in the historic East Armory of the Ninth Massachusetts Regiment, was on that day borne forth, with sad pomp and circumstance, to receive the respectful honors of the people and the last rites of the Church.

It was such a military funeral as even the grand old Church of the Immaculate had never witnessed. Besides the throngs which overflowed and filled the neighboring streets about the sacred edifice, there were Councils of the Knights of Columbus and delegations from the Loyal Legion, from the American Irish Historical Society, and from various civic organizations, come to pay tribute to the heroic dead, and to the bereaved mother and children, who occupied the place of solemn honor close to the sanctuary.

It is not always necessary or useful, in a biographical sketch, to detail the family antecedents of the subject, though there are few biographers who refrain from the recital. But the career of Captain John Drum, for the stirring interest of his long thirty-four years of active service in the army of his adopted land, first as a volunteer in the Civil War and then as an officer in the Regular Army; and for the vigorous influence he undoubtedly exercised over the militant spirit of a son who was destined to be himself a soldier, but in God's army—such a career deserves more than a passing notice. There were many accounts of his life and of the gallant death which closed it, which were printed in the *Michigan Catholic* of Detroit, the Boston *Pilot*, the *Catholic Union and Times* of Buffalo, and other papers and magazines of the time. The following paragraphs are taken from the sketch which appeared in the *Catholic World* for August, 1898, under the title "A Catholic Soldier," by John Jerome Rooney:

> Captain Drum was not alone an example of the highest type of American citizenship, of the American soldier, but he was also the true model of the American Catholic—that type that has added new glories to the Church and is daily imparting strength and beauty of character to the American Republic itself.
>
> The story of his career is therefore more than a story of an individual; the honor that is paid to his memory is more than an honor paid to his heroic life and death. In the largest sense his was a representative career, the manner of his death but the crown and glory of his daily work and his heart's aspirations. He was so true to his duty in every sphere of his long and varied career, that he becomes, in his soldier's death at the post of duty, the noblest type of American Catholic citizenship. This too, during thirty-three years of army life,—life upon the frontier, amid the distractions of camp and fort, far removed from his church and people, and oftentimes in the midst of influences secretly or openly hostile to his faith. Yet all this and more Captain Drum was, not grudgingly, but happily, and by the inevitable force of his manly Catholic character.
>
> During his four years' residence in New York City, to which he was detailed by the government as military instructor in St. Francis Xavier's College, the writer came to know him well; he came to know him as a soldier, filled with the experiences of over a quarter of a century of army life, yet gentle and as open of heart as a child; without pretence of learning, yet more truly cultivated, more extensively read, more acquainted with history and

literature and with the true relation of the world's great affairs than many a man who had spent his entire life in their pursuit. Nor was he a lover of learning for himself alone; in him was conspicuous that beautiful trait of the American Catholic—a life-long determination to bring the blessings of education to his children. As one of his sons remarked: "If we loved him for no other reason, we must have loved him for the great sacrifices he made to give us all the best obtainable Catholic education." One of his sons today is a Jesuit—the Rev. Walter F. Drum—and he had the proud honor of assisting as subdeacon at the Mass for the repose of his father's soul.

The writer was with Captain Drum the night before he left New York to join his regiment at Fort Sill, Oklahoma, just previous to the declaration of war with Spain. Two years ago his time of enlistment in the army had expired, and he might have retired with the rank and pay of a Captain. He was thinking of doing so early this year, but the moment the danger of war grew great he put the thought aside. "The country needs my training,—I shall go." There was the key-note of his character. Nor did he expect ever to return. He showed this premonition in many ways. His last letter to his family, from Siboney, near Santiago, told how he had to his great joy found a priest in one of the regiments and, walking shoulder to shoulder with him on the march, had made his confession. "Pray to God," the letter said, "that I may do my duty." Not a word for his personal safety—his duty was his goal and only wish.

And so, when the charge came, Captain Drum, with his fifty-seven years upon him, and under the tropic sun, was found at the head of his men, leading them on to victory, showing them how a true man can live and if need be, die. He was the only officer in his brigade of the Tenth Infantry who was killed, although all were wounded in that terrible engagement of July 1st and 2nd.

These were the great lines of his career; a few words of the details of his life. He was born in County Cavan, Ireland, May 1, 1840. His elementary education was received in the national school of his birthplace, the town of Killeshandra, and later, from a private tutor, he obtained some knowledge of the classics. At the age of fourteen he came to the United States, remained in New York a few months, and then went to California to seek his fortune, arriving there in 1855. At the outbreak of the Civil War he was deputy United States marshal and was active in the State militia in San Francisco. Volunteering his services for the war, he became a lieutenant in the Eighth California volunteers for about a year and a half. At the close of the war he was elected journal clerk of the California Assembly; but his love was for the army and a military career, and on July 22, 1866, he tried for and obtained a commission as second lieutenant in the regular army. He was assigned to the Fourteenth United States Infantry, and spent two years in Arizona and Southern California, acting as commander of a company of mounted infantrymen for the protection of settlers from the Indians. In December, 1870, he was stationed on the Rio Grande with the Tenth Infantry. For nine years following he served in Texas, participating in campaigns against the Kiowas and Comanches, under General McKenzie in 1874, and later entering Mexico under Lieutenant-Colonel, now General, Shafter at a time when the two countries were on the verge of war. Immediately after this trouble subsided he was ordered to serve with the troops against Geronimo and his band of Apaches, who were forced to surrender to General Miles. In 1887 he was sent on recruiting service, and spent a year in Buffalo and another in Milwaukee. In October, 1889, he returned to Fort Union, where his company was stationed, and was afterwards transferred to Fort Wingate, New Mexico. In January, 1894, he was appointed Military Instructor in St. Francis Xavier's College, New York City, the detail expiring last February. At the time he might have retired, with the consent of the President of the United States, bearing the rank and pay of Captain, but the disaster to the warship Maine at once brought the country to the brink of war, and Captain Drum decided to rejoin his company at Fort Sill. When war was declared he went with his regiment to Mobile, thence to Tampa, and on to the fatal field of Santiago.

Captain Drum was a thorough soldier, in love with his work. General Chaffee, who commanded one of General Shafter's brigades at Santiago, when he was military inspector of the department of Arizona, stated in an official report that Captain Drum's company was the best drilled and disciplined company he had ever inspected. He was ever watchful of the interests of his men, saw that their food and equipment were of the best, and thus, by his care for them, secured their love and co-operation. Powerfully built, five feet eight inches tall, broad of shoulder and deep of chest, he was known among the Navajoe Indians at Fort Wingate as "Thunder Voice." He was a member of the California Commandery of the Loyal Legion and an honorary member of the Army and Navy Union.

Captain Drum was married in San Francisco, February 24, 1868, to Margaret Desmond, of Boston, who with six children, five sons and a daughter, survive him.

When the news of his death was confirmed, a solemn high military Mass of Requiem was celebrated at the Church of St. Francis Xavier, New York City. The Catholic Club attended the services in a body—a tribute of honor never before accorded to a dead member. His battalion of College Cadets were present in uniform to the number of several hundred. In the center of the aisle was a catafalque, covered by the American flag, and upon it rested a Captain's helmet and sword. Then the bugler blew "taps" to the gallant heart resting far away in a soldier's grave. [This was on Wednesday, July 13, 1898. About six weeks later, the remains of the brave soldier were brought from Cuba, and a second funeral ceremony, even more splendid than the first, was celebrated in the Church of the Immaculate Conception, Boston. At both occasions Walter, who was then a Jesuit Scholastic, was subdeacon at the Mass.] Later the body was again disinterred, and conveyed to Washington; and now rests in Arlington Cemetery.

Thus passed all that was earthly of a true American, a true Catholic, a true man. The inspiration of his life and his death cannot pass away. Such souls are a special providence to the world, a grace and a blessing to their kind.

CHAPTER III

THE RAW RECRUIT

Margaret Desmond Drum came of a family distinguished in Irish history. Legendary and historic lore mingles with the name of Desmond, whose bearers claim descent in a direct line from the famous Geraldines. In more recent times, theirs is an enviable distinction in the number of priestly and Religious vocations that have blessed the family and the immediate connections of the Desmonds. A letter from one of these, a Sister of Notre Dame in Wisconsin, to the writer of these pages, gives the names and location of eighteen nuns and fourteen priests who are more or less closely related to the subject of this biography, and adds that there are others also, bankers, lawyers, doctors, teachers, etc., who are near cousins on either the Desmond or the Drum side of the family. Their varied fortunes have scattered them to Canada, the United States, England and Australia, while some of them remained in Ireland and one attained distinction in the missions of Singapore.

As may be noted from the foregoing sketch of the career of Captain Drum, it was inevitable that his family was to see an unusual variety of experiences in all the travels and the constant movement entailed by army life. It required courage, too, and devotedness, of a high order, to brave all the privations and difficulties, to say nothing of the danger, of those long journeys in the days when railroads were as yet scarcely known in the great West.

With regard to the birth of Walter, who was the second of six children, there were certain unusual circumstances which, in the light of later events, may well be taken as indications of a special providence. God seemed to wish to bless the courageous faith of the mother, and to give her a visible proof of the rewards of humble trust in His mercy, and of virtuous loyalty to duty. Mrs. Drum had suffered most severly previously to the birth of her first child, and at one time was perilously near to death's door. Although she recovered from this illness, the physician had predicted that in all probability the next child-bearing would prove fatal. This warning naturally filled her Catholic heart with inexpressible perplexity and terror. No one was near to whom she could turn, in her panic and distress of mind—they were then at a distant outpost, in Arizona—until one day, a lone missionary stopped at the camp; to him she confided her distress and fears.

The good priest gave her comfort and reassurance, and instructed her, in words of faith that were never to be erased from her mind, in the duty that every Christian mother should follow.

After all, he said, we are ever in God's hands. He made us for Himself. He guides our destinies. He has given to each of us a line of duty by which we are to fulfil our destiny in this life. The least we can do is trust Him and do our part faithfully. And if in the doing of our duty there is suffering, and pain, and even more, He will take care of us. He knows best.

Happily, the doctor's prediction proved to be entirely unwarranted, and Mrs. Drum lived to see her children, five sons and one daughter, grow up around her. Shortly after Lieutenant Drum had been detailed for duty at Taylor Barracks, near Louisville, Kentucky, Walter was born, on September 21, 1870. He was baptized under the name of Walter Francis Drum—taking the name Xavier later at confirmation—in the Dominican Church, on Sixth Street, in Louisville. This Church was subsequently destroyed by fire and all the records lost, and when later on, at the time of Walter's application to enter the Society, it was necessary to submit his baptismal register, as is usual in such cases, no other authentic record could be found save that which was made in the handwriting of his father in an old family bible, setting forth that he had been baptized, and giving the names of his godparents. Father Drum used the initials W. F. Drum, or W. F. X. Drum, up to the time of his first vows; he then adopted M (Mary) for his middle initial. Later on, he simply signed himself briefly, W. Drum.

Shortly after Walter's birth, Lieutenant Drum was assigned to the Tenth U. S. Infantry, then stationed at Ringgold Barracks, Brownsville, Texas, and immediately started with his family to report at his new station on the Rio Grande. This meant a long journey, by rail to New Orleans and thence by boat to the mouth of the river. A storm arising at the latter end of the journey added to the distress of the young family, and it was with difficulty that the passengers were disembarked. Most of those on board had suffered severely from seasickness.

Walter's earliest impressions were therefore to be gained amid the bustle and excitement of a Western military outpost, in the midst of Indian tribes and at a distance of over a hundred miles from the nearest railway. His first nine years were spent in such surroundings. How many a youngster nowadays would thrill at the thought of such a boyhood, and envy the lad for all the adventures, which most boys can only read and dream about. And there were adventures a-plenty, and long journeys to be undertaken in company with his father, not only on the Texas plains but also far out to the Sierras and over the mountains to California; and once his father's duty called him eastward as far as Boston. This latter trip, however, was in 1872, and therefore not likely to be remembered, but later journeys were full of incidents that impressed themselves on Walter's young mind.

The vast regions of the West were as yet but sparsely settled, and journeys in those days were far from comfortable and often very dangerous. One of these long journeys is thus described, as it fixed itself on his retentive memory.

> It was in Texas down by the Rio Grande, we were two hundred and fifty miles from Houston, the nearest railroad station, and traveled overland in what we used to call a prairie-schooner, a lumbering wagon that took us two weeks to reach the railroad. All one day we found no water. I was a child, about seven years of age at the time, and I remember the thirst to this day. At last we reached what was then called the "Buffalo Licks." They were mere mud-holes where the rain had gathered, and where the buffaloes had kicked up the mud while licking up the water. Stagnant and filthy as this water was, we gathered in as much as we could, and boiled it for drinking. How foolish it would have been to have taken that filthy water to drink, if nearby we had found a font of living water, a spring! Yet that is what we do when we leave the Font of Living Water and turn to sin, and

make of our soul a mud-hole, a "Buffalo Lick" of stagnant waters.

The following is a fishing story of those boyhood days that was frequently used in later retreats;

> Have you ever gone trout fishing in the Catskills? I remember trout fishing in the Rockies, years and years ago. We were four or five boys in the party on this expedition. At first it was vain. I cast a fly; the trout made for it, turned tail and darted away in disgust. I tried all manner of bait. Not attractive enough! They were refined, educated fish. Suddenly, down darted a grasshopper. And up darted a little trout. Into the air it leaped and upon the unwary grasshopper it pounced. Then I said to myself; "These trout are epicures; they are too well-bred; they scorn the ordinary coarse food; they must have dainties served up to them and since they show such a marked preference for grasshoppers, why, grasshoppers they should have. That day I caught 40 trout, and in two weeks we had 960! That is no fish-story!
>
> Rather earlier than this—it was in the Little Rapids of St. Mary's River, between Lakes Superior and Huron. The trout there was more game. There we had real sport. What a thrilling pleasure it was to see the trout leap from the water and pounce upon the bait; to play with him; to reel in and let out the line; to bring the little spotted beauty within reach. But it was not easy to do. There in Northern Michigan it was quite a joy to bring home only one or two of the glistening, speckled rascals as trophies of our sport.
>
> Like fishes are caught in a naughty net, so are the sons of men caught by Satan in an evil hour, whensoever he cometh upon them unawares. The devil will fish a long, long time for the game fish, and will be satisfied to catch but one game fish after a long, long time of fishing."

One rather dangerous experience, which, however, displayed the lad's presence of mind in a remarkable way, was once described in an exposition of the passage from Proverbs xxx, 18; "The way of the eagle in the air, the way of the snake on the rock."

> Because the snake wiggles on the rock, and leaves no mark, the action of the snake is hard to understand unless one has specialized in snake-ology. That action of the snake never interested me except years ago when a rattlesnake actually did walk over me, probably took

> me for the rock. I was a little lad about seven years of age at the time, and in a green, rocky cave in which several of us little boys were playing, a rattler was suddenly aroused. As he made towards me, I threw myself on my back. He just wiggled over me, took me for the rock.

Once, as a Scholastic at Gonzaga College, Washington, D. C., Mr. Dum gave his altar-boys a talk on vocation, taking his cue from a hunting trip in the Rockies; a part of this talk is of interest at this point:

> A few years ago, I was with my playmates a-hunting quail, in Western Texas and near the Mexican border, down by the Rio Grande. We were only boys, the oldest fifteen, the youngest ten. Fearless we wandered through parched valleys, climbed many a steep crag of the Rockies, started the rattler a-going, knocked down a few jack rabbits and bagged a few quail. We had a jolly good time all day long, till some one discovered that the little ten-year old chap was gone. How we hunted, and shouted for him! It wasn't a bit of use. The echoes came back from the cliffs hard by, but never a trace had we of the youngster we had lost. Sad of heart we returned home, to tell the sorry news. And what do you think? The little fellow was there before us!
>
> We were then living by the bank of the Rio Grande, and the youngster knew that the morning sun always shone in front of the river; so when he got lost, instead of sitting down to cry, he began to walk straight in the direction of the sun, until he came to the river, and then walked down the river and thus struck home.
>
> The only thing that will lead you straight home is to keep right on with the sun. That is the sun of grace, and it is your duty, in the bright light, in the dim light, or even in the darkness, to go always towards the sun that will lead you to Heaven, your true home. You must pray, and feel sure that unless you follow God's grace, if you do not walk straight towards the sun, you will never reach the Rio Grande, the Great River, the state of life down which you must journey in order to reach Heaven, your home.

During these earlier years, naturally, there was little opportunity for schooling, out there on the frontiers, though we have no means of conjecturing just how much a red-blooded youngster would seriously regret that omission. In the case of Walter and his brothers, the loss was made up to a considerable extent by the personal tutoring of his father, who had some knowledge of the classics and was well-grounded in mathematics, or of some other soldier who would occasionally be found to conduct a school for the children of the garrison.

Years later, on a visit to San Francisco, Father Drum recalled that the first regular school he ever attended was in that city. It was referred to in his great sermon on Socialism and the evils of the day, as reported in the San Francisco *Chronicle*:

> I spent my first two months in a public school in San Francisco. It was the first school I ever attended. I learned to spell by phonetics there. Then I went to a parish school and was taught to read and spell.

In May of the year 1879, the Tenth Infantry was transferred to the Lakes Station, as it was called, and the company to which Lieutenant Drum was attached, Co. I, was assigned to Fort Brady, at Sault Sainte Marie, Michigan.

For the first few months the boys were placed in the public school at the "Soo," as the city was then and is yet called. Lieutenant Drum had not as yet come in contact with the Catholic school-system, and had formed for himself very decided views on the advantages of a public-school education. Even Mrs. Drum's constant urging had hitherto failed to convince him of the advantages of Catholic education. A slight friction, however, that resulted from the refusal of the Principal to allow Walter to recite a little Irish poem at a school entertainment, led him to suspect that the atmosphere was not altogether a suitable one for the growth of his sturdy Catholic plants. Walter had committed to memory a pathetic little ballad of the days of the Irish famine, entitled, "Three Little Grains of Corn," and although he was able to recite it creditably, the Principal sent him back home with the injunction to learn something else that was not "so pious." Lieutenant Drum was thoroughly indignant and came himself to the school to demand an explanation, and when he perceived that there was no other reason for the objection than that of mere bigotry, insisted that the child be allowed to recite that poem and nothing else, or else to withdraw his name from the program. Walter recited "Three Little Grains of Corn," in a tender little soprano that brought tears to the eyes of the audience.

But Lieutenant Drum thereafter began to entertain grave doubts as to the advisability of keeping his young children in such evidently un-Catholic surroundings. It was therefore fortunate that about this time the family came under the influence of Father Chartier, S.J., who was then pastor at the "Soo" and had previously been stationed at Fordham College, N. Y., for some years prior to the setting-off of the Canadian province. On Father Chartier's advice, Lieutenant Drum at last gave his consent to have his children placed in the parish school.

It is not easy to foretell the man that is to be, in the few letters that a boy can be induced to write; but there are in the following selections characteristics that reveal the same thoroughgoing personality and the same earnestness that was to mark Father Drum through life. His brother, John Desmond Drum, known in the family as Des, or Dessie, was at Assumption College, Sandwich, Ontario. Walter writes to him:

Saulte Ste. Marie,
Nov. 1, 1883.

Dear Brother Des,

I did not finish the letter that I wrote to you, on account of Papa sending it away while I was at school. We have a new teacher and we have compositions, one every week. Our first was on wealth, 2d on the fourth commandment, 3d on flowers. I am in the Progressive Higher Arithmetic. I could have commenced Algebra but I preferred the Arithmetic. How are you getting along in your studies and especially in Geometry and Algebra? I am getting along pretty well in my studies When you went away we just had about $5.00, and we have about $8.00 in the bank. I paid 10 cents the other day to go to a negro show called the jubilee singers. I am fixing up my stamps a little better, if you can tell me of some stamps you want me to send you I will try and send them to you. I traded a New South Wales stamp for a Mexican stamp, and a Canadian Bill stamp for two U. S. Bank Note stamps, from Will Kirkman. Yesterday was holy eve, and we had lots of fun. Will Maher and some other boys tore down some gates and did a lot of mischief; did you have any fun on that day? They tore down the target butt a few weeks ago, if it wasn't for me having to go to school I would have got about 150 lbs. I only got 55 lbs. and sold it for $1.32 to Mr. J. B. Eddy.

Andrew F . . . . was drowned last Sunday in the canal while he was drunk. Are there many new pupils; Fred Sevald said he knew that his cousin was at the college, and maybe he would go. From your loving brother, W. F. S. Drum.

Sault Ste. Marie,
Dec. 5, 1883.

Dear Brother Des,

I received your letter dated Nov. 30. Did you receive the letter that I sent you dated Dec. 1? I hope you are getting along well in your studies. Oliver Saunderville is very sick. We got a letter from Aunt Josie, her and uncle Everette are going on a trip on Dec. 10. Uncle Everette caught 9 whales. We are all well at home. I love to write to you. The new teacher is the same one that was teaching when you were here, her name is not Miss La Pierre, but Miss St. Pierre.

The rest of the letter gives a long list of the kinds of stamps which he had in his collection. Every schoolboy was a philatelist in those days, and Walter had several books in his stamp collection. But at the end of this letter appears a stern order in the handwriting of his father, telling John not to send any more stamps to Walter, as he was spending too much time with stamp books and neglecting his studies.

After the Christmas holidays he writes again to his brother,

Sault Ste. Marie,
Jan. 10, 1884.

Dear Brother Des,

I wish you would write oftener, for we were very anxious to hear from you. Did you receive the letter that Joe and I wrote to you and that I directed? Please send me lots of stamps and gather as many as you can for me . . . I spent a pleasant Christmas and the small children got lots of presents. I got three prizes at the brother's affair, 1st prize for singing, 2d prize for regularity, and 3d prize for coming to funerals [he was an altar-boy at this time] and about 15 other articles I got from brother for certain things, most of them I gave to Hugh please write soon. I don't thing you can read this letter as I wrote it fast remember to send me lots of stamps [he had not seen his father's stern injunction to John] and tell me if you want any more . . .

The letter thus rambles on without a thought of punctuation or a pause for breath.

It is interesting to note that the excellent work of this Catholic school, and the influence of Father Chartier himself, very soon converted Lieutenant Drum into an ardent champion of Catholic education, and there was never again a question of permitting any of his children to attend the public school, if a Catholic school were available.

Walter spent about four years in the parish school at the "Soo," and grew to love his teachers with a devotedness that remained with him through life. He often alluded to this period of his boyhood and to the wonderful influence that was exerted upon his growing mind by these devoted Catholic educators. One of his teachers was the saintly Miss Odilia St. Pierre, mentioned above, who remained his closest friend of later years, up to the day of her holy death as Principal of the Nardin Academy in Buffalo. Walter also learned to serve at the altar, and thus came into closer contact with Father Chartier himself, as well as with a cheerful old lay-brother, who is spoken of in the last letter—Brother Fountain by name—who was in charge of the sacristy, and with whom Walter became a special favorite because of his dependableness and fidelity.

> I can never forget [said Father Drum in a later retreat] that one of the most helpful influences throughout my life has been the affection of the devoted women who taught me in grammar school. I am very glad that they taught boys of my age, so that although I was supposed to have finished the grammar school at thirteen, I was allowed to stay with them till I was fifteen, because of the rather haphazard schooling that I had received before coming to the Soo. A boy of fourteen or fifteen becomes very devoted to a good teacher. I never forgot the influence of the last teacher I had there. It was like the influence of my mother . . . My post-graduate course in this Catholic grammar school was training! I must have got some good out of it, for I afterwards managed to finish the whole course of high school and college in five years.

The last assertion is a statement of actual fact, although it is now a difficult matter to follow the record of his full course through high school and college. It must be remembered that the courses were not so definitely marked out in those days as they are now. Moreover, Walter was several times allowed to be promoted twice during the same year. His father's military movements made it necessary to take his children with him, and it thus came about that Walter saw as many as seven or eight schools and colleges in different cities; and somehow it was easier then than it is now to get oneself put in a higher class.

But it is certain that even at this early age Walter began to show his unusual intellectual powers. In spite of his previous handicaps he soon finished the work of the grades, and under Miss St. Pierre's capable personal direction, made rapid progress in the fundamental high-school subjects. His voice, too, had early developed into a fine boy soprano; and when the lad began to take singing and piano lessons, he manifested a musical talent which might have led to an entirely different career for the future Father Drum. But his interest in music, at least in these early days, was confined to singing in the church and school choirs, and to occasional vocal solos at school concerts.

One of his brothers remembers from this period an instance of remarkable pluck in the slight, frail lad, who once stood up to defend his smaller brother against the bullying of a much larger boy; "It was the only fight," he said, "I ever saw Walter engaged in, and although he received a grueling punishment, it was the greatest exhibition of calm courage, gameness and sturdy persistency I ever saw, and we army boys, you may believe, had to fight every inch of the way."

In 1884, the regiment again received orders to march. This time its field of operations was to be in the very heart of New Mexico. The children, therefore, were obliged to give up their pleasant school associations at the "Soo," and, very soon after reaching New Mexico, it was arranged that they should attend the post or garrison school at Fort Union, the only one available. But a few months later the family moved to Fort Bliss. In the El Paso school, Walter was put in the highest class, which was really a class that corresponded to a first year of high school, while his elder brother John attended the college, conducted by the Jesuits, at Las Vegas. The remaining children, Mary, Joseph and Alphonsus, were in grammar school with Walter.

The following are a few typical letters from one American boy to another:

Fort Bliss, Texas,
January 11, 1885.

My dear Brother,

I will try and not be so careless after this, and write oftener. I am in a base-ball club (called the D's) we wear a regular uniform. I play short stop and left field. I am in the highest class in the school at El Paso, one of the boys in my class is the captain of my club. Henry Barrette and Hamilton Hawkins are in another (called the L's) the first time we played we beat them by 10 tallies, the last time we played they beat us by two tallies. I got acquainted with Johnny Douglas he told me all about the college, and took some things to you. Enclosed find list of stamps. I have lots of coins now, and am trading stamps for coins. Papa is shooting at 500 yards, figure of merit of Co. H is forty. The 2nd Lieut. has come. His name is Wren. I am in the highest class in the school. Joe is in the next, Allie is in the same class as Henry Barrett which is the 5th. I am in the fractions of Algebra. I have not studied neither Latin nor Greek since I left Union. Try and get some good stamps for the War Dept. as they are getting scarce. Hoping to hear from you soon I remain your brother that will write to you often.

Walter Drum.

Fort Union, N. M.
Oct. 20, '84.

Dear Brother,

Mama has one of those sick headaches today, so I write a few lines for her. Miss . . . went to Las Vegas today, but as Mary has left us we could not send anything to you, not even a letter . . . I found the skate key, and will give it to you at the station. I went up to the adjectives in Greek, and commenced in the beginning of the Latin grammar and finished the verbs, I do not study any more than that. Lieut. Clay was so far ahead in the Army shooting that it was not possible to be beaten, but on the third day he heard of his brother's death it shocked him so he only made 70 some out of possible 105 and only took second prize. Papa will despatch to you when we leave . . . Be sure to be at the depot in time and be ready to answer lots of questions. Mama says if you have any thing to say to answer this letter as soon as you get it. We will not leave till the 23 on the evening train.

Yours truly Walter Drum.

P. S. Mama met the Father Prefect he spoke a good word for you.

Walter F. X.

Fort Bliss, El Paso, Tex.,
June 14, '85.

My dear Brother,

I received your letter of the 24 of May. Mamma sent some oranges to you through Mrs. Harris today. Our examinations will commence on the 14th of this month and end on the 19th. If I pass I will be supposed to be a graduate in Reading, Spelling, Arithmetic, Algebra, Geography, Physical Geography, Civil Government, Grammar, U. S. History, and Physiology; and the books I will study next year will be Geometry, History of the World, Physics, Latin and Rhetoric. I have 6 pigeons. Pretty hot here. The Apaches have broken loose and are going towards Mexico, the Navahoes and Musculares are on the war path. England and Russia have not commenced war yet. Gen. Grant is still sick. I hardly think that I will continue to the lessons on the violin as Papa says I will be going with you next year maybe. We won't receive any premiums at the end of the year. I hope you get some good ones, but don't study too hard. We received your monthly reports . . . We found a place with a good bottom and I am going to try to learn how to swim. We have had 37 little chickens hatched but 12 got killed. I had 8 pigeons but 2 flew away.

Your obedient servant, your brother and pard—
W. F. X. Drum.

The regiment itself was stationed at Fort Union, some twenty miles distant from Las Vegas; but in September of the year 1884, Lieutenant Drum was promoted to a captaincy and was assigned to Company H at Fort Bliss, near El Paso. Here he remained till February, 1887, when Company H was again transferred back to Fort Union. These constant changes made it difficult to arrange for the schooling of all the children, but Walter was entered with his brother John at Las Vegas College in the Fall of 1885, remaining there till the end of the year. The family remained for the most part at Fort

Bliss, and there received frequent letters from Walter and John, some of which, like the following, were written conjointly:

Las Vegas College
Las Vegas, N. M.
Oct. 14, 1885.

Dear Parents,

Walter and I have written two letters which you have not answered; but we write again in order to send our monthly reports. I hope you will be pleased with mine. It is the best I have ever received. I have not lost a mark in anything yet. Walter's I think is the second best in the college, but he has to study pretty hard. He is making good progress in Latin. I have had a cold all during the last week . . . I was elected President of the Athletic Society. Walter and I serve Mass every morning except Sunday and Thursday. Father Lezzi the teacher of singing is a very fine musician. He has advised me not to sing any more for a few months, as my voice is now changing and if I continue I may ruin it. As the target year is now over and papa is not so busy I hope he will write us a long letter and tell us all about it . . . On account of being so sick you must excuse the bad manner in which this letter is written.

Your loving son, J. D. Drum.

On the back page of this letter Walter wrote to his sister:

Las Vegas, N. M.
Oct. 18, 1885.

Dear Sister, Mary.

You dont know how lonesome it is here without hearing from home. Please write to us. Are you getting along all right in your studies Are Joseph and Allie getting on the same? Does Hugh know his ABC's yet? Our reports will be sent with this letter, I have two 99's but will try and do like Dessie did this month. I belong to Dessie's nine of baseball, tell Joseph, that we beat the other side by three runs. We were playing for twelve bottles of soda water. I belong to the Debating Society. Does Florie and Aleck go to school yet? . . . Hoping very soon to hear from Allie, Joseph, Hugh and yourself,

I remain, Your loving brother, W. Drum.

Las Vegas College
Feb. 6, '86

Dear Joseph,

We received your letter of the 19th ult., and were glad to hear that you are well. Desmond wished to write but I did also, so it was decided in my favor do not think it mean of Desmond. Why don't you write to us oftener and tell us how you and the children are getting along. Mary is getting along well I am sure because she always got good marks at Ft. Brady . . . We have finished our examination, and are in the second term now. The days are passing slowly but surely, and in a few months we will be on our way home. For my mark in examination I got 99, which was pretty good for a lazy fellow like me. Desmond got 100. We are now studying Phaedri Fabularum, or Aesop's Fables in Latin verse. It is very hard for me to translate on account of the mixture of the words such as having the adjectives about three lines away from the noun with which it agrees.

Desmond will answer Papa's letter. He is studying Horace in Latin and Xenophon's Anabasis in Greek.

Why you are getting to be quite a hunter Joe killing four meadow larks is better luck than ever I had. Why can you not get more pigeons? Are there not lots of them around the fort?

Make Allie write to us it does not matter if he does not write good his letters are always very interesting and make us laugh. He only wrote one letter to us since we bid him goodbye.

With love to Uncle, Aunty, Mother, Father, brothers, Sister, and all my friends,

I remain Your affectionate brother,

W. F. Drum.

It appears that both Walter and John remained at home during the first half of the school year 1886-1887, on account of the latter's illness, and after recovery were tutored for a while by a Father Di Palme, S.J., who prepared Walter to enter the second term of Fourth Year High or Third Grammar as it was then called, while John was coached for the class of Humanities. Both were again enrolled at Las Vegas after the mid-year examinations.

A new order came from the War Department in the Fall of this year (1887) assigning Captain Drum on a recruiting detail in the East. This was a most welcome opportunity for which he had long been hoping, for it meant a station in some large city where he could give his children ampler opportunities for educational advancement. Captain Drum therefore came on with his family to Massachusetts, to visit relatives in Boston, before the definite assignment should come from Washington.

But when he reported for duty a great disappointment was awaiting him. The War Department had decided, on account of the depleted strength of the army, to retain on recruiting service the same detail of officers that ordinarily would have returned to their regiments. The result of this new arrangement was that officers of the new detail, like Captain Drum, were again obliged to go west. He found himself assigned to the district of Minneapolis, in Minnesota.

However, Captain Drum decided to leave his family in Boston till he could investigate the school facilities at his new post. The boys were therefore entered at Boston College, and according to the college catalogue for the year 1887-1888 John was assigned to Humanities, although he believed himself entitled to the class of Rhetoric, while Walter was enrolled in the class two years below him, Second Grammar as it was then called. John remained at Boston College, but Walter did not stay to finish out the year. His father, after being a month in Minneapolis, was offered an opportunity to transfer to Milwaukee, which he quickly accepted as soon as he learned that there was a Jesuit college in that city. He therefore sent for his family from Boston, and all but John came on to Milwaukee. For some unexplained reason Walter was enrolled in the Freshman class, at Marquette College, and at the end of the year finished successfully, but only by dint of grueling study.

> Morning after morning, [wrote one of his brothers] during the terrific lake region blizzards, I have seen him strap his books, go to the cathedral and serve his Mass, and then trudge through snowdrifts for two miles, to be on time for his classes . . . Had he told us anything of his purpose, things would have been made easier for him. But he took his medicine with the rest of a large family the parents of which had but one objective, a Catholic college education for each and everyone of the flock. When one considers that this was finally accomplished on an army officer's income which did not average more than $2500 a year, the creditable feat may be appreciated. And during those years the army orders took us on thousands of miles movements at great personal expense—the government did not pay the travelling expenses of its

officers' families . . . I believe Walter had made up his mind for the priesthood all along, but there was never a word spoken regarding vocation.

These educational wanderings were not yet over. In the Fall of 1888 Captain Drum was assigned to recruiting duty in Buffalo, and brought his family to that city. Walter, as was his wont, had continued to study all during that summer, and when at the opening of schools went to be enrolled at Canisius College, succeeded in having himself put down for the class of Junior! However he managed the maneuver, he found it necessary to study at a tremendous pace, but he was determined to make good in the class. His brother says that he started the year with a handicap of zero for his marks in September, when the family were on the move, and yet in spite of this misfortune he managed to finish the year with an average of nearly 80 per cent. It was probably at this period of his career that Walter developed that capacity for tremendous hard work and incessant study that was to mark the rest of his life.

As a young man, Walter's physique was never very rugged, and his parents had often to worry a great deal over his physical condition. This was especially true during his year at Canisius. He had never shown much inclination or aptitude for athletic sports as such, although he was not averse to exercise and took part in the usual games of boys of his age. But at about this time he began to take no part whatever in athletics, of any kind, devoting himself entirely to study and to reading in order to keep up with his class. It was burning the midnight oil in a quite literal sense. Argument to the contrary had no effect on him. His brother Joseph writes:

> I have seen the light in his room burn until four o'clock in the morning. Gradually he seemed to isolate himself in his intense studies. So much did he exclude himself from us, that in John's absence at Boston College, the responsibility of oldest boy at home fell upon me . . . Time and time again I went to his room wondering, worrying about him. Father would issue his order to Walter to go to bed at a certain time. He would smile, go to his room and lock the door. Then he would study, study, study. On his desk he had a little tin can cover, the sort in which friction primers came. It was kept filled with cold water.

As his tired eyes would droop he would moisten his fingers and rub the water on the lids to keep awake. Mother might knock at the door. No answer. Father never interfered except again to issue his order, knowing it would not be obeyed. When one remembers his own grim efforts to gain an eduaction, by the light of his miner's lamp in the gold mines of California, we can understand his secret satisfaction and his sympathy with Walter's devotion to his books.

Time and again mother would appeal to me to help her stop him. I tried every means of a thoroughly inventive mind. I sang and whistled outside his door. No answer. I played tick-tack on his window, and threw pebbles at it—all in the effort to arouse that temper which I knew was smoldering. But it was a new Walter. He paid no attention to me. And yet through it all his parents never infringed on his personal liberty by taking his key away. That would have been unforgivable in the minds of all of us.

At this time Walter began to read very extensively in order to cultivate a literary taste, and also showed some talent for writing. He once said that during his Junior year in college he had read through the entire Bible. The only recreations he allowed himself was an occasional performance at the theater, always choosing the higher-class performances; and he took to making frequent visits to the Grosvenor Public Library, not to read, but for the sake of meeting an old gentleman who was always sure to be found there, and whom he would challenge to a game of chess. The gentleman happened to be Buffalo's champion chess-player; but although Walter did not often win, a couple of theater-passes would always be his reward for his courtesy. There was also a rare trip to the Falls of Niagra and to the Canadian shore. His first visit to the Falls left him vivid impressions, some of which he recalled long afterwards, in a retreat, in connection with the attractiveness that the whirlpool of the world may sometimes exert over those who have left it:

Think of the risks men take, and often for no other reason than the risk! I shall never forget the fright such fascination once caused me in boyhood. While bending over the deal railing and peering over into the abyss at the American Falls of Niagra, the inclination suddenly seized me to plunge in. I shrank back in horror. The memory of that boyish terror is with me still. Such terror should we feel at the fascination of the plunge into the abyss of worldly pleasures.

About the end of this year Walter learned that he was to return to Boston in the Fall. It was his mother's wish that all the boys should finish at Boston College. Walter was apprehensive of the assignment that would be given him. Only two years before he had been in the High School, and it would be difficult to convince the authorities there that he was now ready to enter the Senior class. Fortunately, an opportunity presented itself to forestall the difficulty before he should reach Boston at all. The maneuver is thus described by one of his brothers:

> About Christmas time the beloved Father Fulton, who was then President of Boston College, visited Canisius. Walter called on him and explained that he expected to go to Boston College in the following year and wished to know what were his chances for entering the Senior Class. Father Fulton questioned him as to his present standing and studies, and not knowing that he had previously been a student at B. C., in a much lower class, informed him that if he would do some special work in calculus and pass an examination in that study, he would be given Senior. This promise was all that Walter desired. The following summer he took up calculus, with my assistance, and early in the Fall took a special examination, under Fr. William Mullan, who notified him shortly afterwards that he had passed easily. Walter immediately sought out Father Fulton and amused him with the announcement: "Father, I have crossed the Rubicon." Later, when Father Fulton found out the full facts about his previous rating at B.C., he could only chide him for the march that he had stolen. But Walter stayed in Senior, and thus attained his wish of many years, to catch up with his brother John; and received his A.B. degree in the college which, three years before, he had entered in the class of Fourth Year High!

CHAPTER IV

VOCATION AND NOVITIATE

The family had believed all along that the priesthood was Walter's one ambition, although, as a matter of fact, there was seldom any open reference to his plans; and he himself neither expressly indicated his intentions, nor seemed conscious of the family opinion in regard to his future. He was always a serious student, rather frail in health, earnest of manner and faithful to his duties, although he was not considered pious in any sentimental sense.

It was, therefore, a cause of not a little surprise and apprehension in his relatives to see the marked change in Walter's habits and deportment during his Senior year at Boston College. His brother John recalls that about this time he became very much interested in outside activities and studied only enough to keep along with his classes. Indeed, he entered so zestfully into social affairs and in general appeared so changed from his former serious disposition that the family were led to fear he had abandoned altogether the idea of becoming a priest. However, his brother conjectured that he had practically made up his mind to enter the Society, and knowing that he could get a thorough course of philosophy later on, decided to conserve his energies to some extent and give himself a little more freedom before leaving the world for good.

Where formerly he was somewhat indifferent as to his personal appearance, during his last year he developed a taste for stylish clothes, carried a cane, and appeared always dandified and trim. He affected an air of dignity and a grand manner that somehow clung to him from this time on, and became the most noticeable mannerism of the later Father Drum. He entered zestfully into the dramatic productions of the students of Boston College. He began to cultivate a wider circle of friendship, and to attend social affairs quite frequently, and enrolled himself as a member of several social and literary organizations about town. On the evening before his departure for Frederick, he attended a party with a large crowd of friends and returned home at a very late hour that night.

One of his activities of this year was to attend a night school where he took a course in English literature. The instructor professed himself an atheist, and often took occasion to exploit his views, or lack of them, on questions of philosophy and religion. There was too much of the Irish in Walter to listen tamely, and it was inevitable that he would frequently engage his professor in warm discussion.

He may have believed that this literary course was necessary in order to enable him to qualify as a member of the Phillips Brooks Club which he joined during this year. Many of its members were Harvard students, and with one of these men Walter became very friendly. The result of frequent conversation with this friend led him to entertain a strong inclination to take up a post-graduate course in English literature, at Harvard, before applying for entrance into the Society. In later life he often said that he thanked God he did not follow out this intention.

The truth appears to be that Walter was all this time fighting the battle of his life. More and more insistently the call to higher things had been making itself felt in his soul; but he fled from it almost in fear. It is quite certain that he went through a considerable struggle with regard to his vocation, and all these external activities into which he threw himself were but the reaction of nature against grace. Many a night he returned home to find himself depressed and gloomy after having, in the earlier hours, enjoyed himself immensely—or so he believed—at some social gathering or theater party. His own words, written many years later, best describes his feelings at this time and reveal the fact that he never forgot the severity of the struggle he had to undergo.

> After every glorious time, the thought would always recur: "Well, that is done, and what is the use of it? That is done, and tomorrow, something else,—and what will be the use of that? Why not make life worth living, and measure it by eternal values?" And so I found it necessary, in order to be able to fall asleep with some peace of conscience, to take up the Imitation of Christ and read a chapter or so, to fall asleep with some ideas that brought in God and made life worth living.

Occasionally he would stroll out alone to the solitude of Boston Common—which in those days was a much more secluded retreat than it is now—and, with only his cigar for company, he would weigh the great question over and over again from every angle. "Smoking stimulates me when I am calm, and calms me when I am stimulated," was a saying of his in after years; and on one occasion he whimsically explained that the final decision to follow his vocation grew clear in his mind amid the encircling smoke about his head. "But," he continued, "this method of meditation has never been recommended in any spiritual book."

Father Drum later grew accustomed to look back on this period of painful uncertainty with a feeling of gratitude, and came to think that the very difficulty he encountered was a comforting proof of the reality of his vocation to enter the Society of Jesus.

> One of the most certain signs of a call to religion [he said in a retreat to seculars at Marymount] is the resentment of the natural feelings. No one could have hated the idea of entering religion more than I did, with my feelings. To me it was simply discouraging to give up all ambition of self-glory; to give up the prospects that had been growing in my imagination, the prospects of worldly fame, and of following in my father's footsteps—to give it all up and knuckle down to obedience, to wear a thing like this cassock. The whole idea of such a vocation did not appeal to me at all. Quite the contrary, my feelings were all the other way.

But yet, the thought of the priesthood had been present to his mind, more or less distinctly, ever since that day in early boyhood when " Father John" first visited the camp at old Fort McKavett, in Texas.

He tells the story in an offhand, casual manner in one of his retreats at Moylan, Pa.:

> The "guest of one day that passeth by" reminds me of the first priest I remember to have met. It was on the plains of Texas, something like a hundred miles from the railway, in the days when we could hear Mass only once during two years. I was a lad of seven. Father John came our way, and I recollect his taking me on his knee and hearing me recite my prayers. And,

> strange as it may seem, the desire to be a priest was then first implanted in my soul. It is "as the remembrance of a guest of one day that passeth by."
>
> However [he continues], it is true no one ever spoke to me about becoming a priest. I think I should have resented it very much if anyone had. I think no one ever suspected such a thing. The idea had been with me since childhood, and could not be got rid of.

Meanwhile, the annual retreat for the college students came around. He had yet reached no definite decision. He always believed subsequently, that his confessor was the unwitting cause of this hesitation, for he had urged him on as if his vocation were a foregone conclusion; so that at one time Walter was strongly tempted to give up the idea altogether, merely because of this Father's insistence.

> During two weeks we had some arguments together on the subject. He urged me; he made a mistake in doing this, though he knew me very well. By the time the Father Provincial arrived in Boston, I had decided that I was going to Harvard to take two years and get a Ph.D. in English literature. I thought that in any event, I could then enter the Society. It would be a great honor to the Society to enter from Harvard. Of course the devil was hoodwinking me. I did wish to have two years more of life, to drink life to the dregs, and then leave the dregs. Two years of it! I thought that after two years of it I could give up the dregs. So this dear, good Priest, who is now dead, met me one day and we had it out again, and finally he said: "Well, you ought to come to Harrison Ave. and meet Father Campbell anyhow. No need to talk about vocation. He knows several army friends of your father." Well, I could not refuse that, so I arrived.

It appears that Father Campbell won his complete confidence and after listening to his plan with sympathetic interest, gave his advice briefly and in no uncertain terms. "Don't be foolish," he told Walter, "you are old enough to make your decision now. Either you are going to be a Jesuit, or not. If you are going to be a Jesuit, enter the Society now, and don't waste two years at Harvard. If you are not going to be a Jesuit, take up your life career at once, and don't waste two years at Harvard." The effect of this advice was immediate and is thus described:

So I went home and wrote my letter of application at once. He sent me to four examiners, and shortly afterwards I was accepted. But I told nobody about it until about ten days or two weeks before the end. One day I came home and found one of the examiners there. I took in the scene at once. It was clear that the good news had arrived. My mother and sister were in tears, and my father looked serious. My fate seemed settled.

Captain Drum, in talking it over with another of his sons, said simply: "It's a soldier's job, Joe—and he can do it!"

His brother John adds another version of the events at this time:

The knocking of his conscience, and the advice of Father Buckley and of the Provincial prevented him from delaying his entrance into the Society.

As I stated above, I had concluded that he intended to remain in the world, and was very much surprised when, early in August, 1890, he confided to me that he had applied for admission and was accepted. We were on a little vacation together; he had already confided his intentions to his parents and received their blessing.

I accompanied him as far as New York. At that time he was a heavy smoker. He had earned money in various ways, like writing space on the Boston *Herald*, which enabled him to enjoy some of the pleasures of life without taxing his parents. He had also won some money prizes at Boston College the day of his graduation. As I recall it, he had a good time on his trip down to Frederick, Md.

On September 7, 1890, Walter was admitted to the Novitiate at Frederick, Md. A concise account of his Noviceship days is given in the words of his Superior and Master of Novices, Father John H. O'Rourke, S.J., who writes:

I think I can truly say from the day he entered the Novitiate, till the day he left for Woodstock after his Juniorate, Walter Drum was more than ordinarily satisfactory and edifying. The end of the Long Retreat found him deeply impressed, for he had caught on to the spirit of the Kingdom, and in that spirit he grew stronger and more generous daily until Our Lord called him to his reward.

During the Novitiate he, at times, suffered from some disorder of the stomach, which may have been the beginning of the malady which has at last deprived the Province of his example, labor, and remarkable scholarship. I remember distinctly that in physical pain he was no coward, and took it not only courageously but even with willingness and joy.

From the early days of the Novitiate the crucifix meant much to him, and he was then and always later ready to follow its lessons. As a Novice he showed a remarkable intelligence and love of the Exercises, which he studied diligently and constantly. It was from this study in the years of his Novitiate that there grew up in his soul that warm and personal enthusiastic love of Our Lord so characteristic of our spirit, so surely detected in Father Drum's life, by those who knew him intimately. The love of Our Lord became the main spring of his personal piety, and gave him the courage to lead a life of constant and strenuous abnegation which characterized him in those days, in fact, in all his days.

He had faults, of course; we all have, but in his efforts towards perfection, his ideals were ever high and his strivings generous and constant. His sacrifices were many and willingly made. He was not of those who must be urged and stimulated to renewed and unflagging efforts for spiritual perfection, but rather of those who have to be kept in check and restrained. He needed a curbed bit rather than the goad and the whip. He was no laggard, no self-seeker, not one of low ideals, of halting or hesitating correspondence to grace, but was rather of the Xavier-type, too generous to shirk or refuse any call to higher things. There habitually echoed through his heart the cry of the Apostle of the Indies, to whom he had a great devotion: *Amplius, Domine, amplius.*

I was rather intimate with Father Drum through all his years in the Society, and I could not but note that this spirit of enthusiastic generosity remained and even increased with him to the end. Another trait I remember in the Novitiate was his constant application to study and self-improvement. While neglecting no duty of his daily routine, and ever intent on his spiritual progress, he read carefully and thoughtfully, the best of the Novices' library, and scrupulously husbanded his time so that when called to Woodstock after one year of Juniorate we felt that he had availed himself up to date of the opportunities furnished by the Society for the spiritual and intellectual formation of her children. This trait carried out in his after-life made him a man of ripe scholarship which reflected credit upon the Society and the Church.

The long life of the Religious Novitiate is of interest to a work-a-day world only as a curious phenomenon that reveals the power of grace and of supernatural things in the lives of high-minded youth. In the zest of sacrifice, earthly ambition is cast aside, and only one purpose, one desire, appears to dominate both heart and mind. The youthful recruit wishes to follow the call of his Leader and King, to try on the equipment of that spiritual warfare to which he feels himself summoned, and by the various exercises of the Novitiate to test his own ability and in a measure acquire some expertness and dexterity in the use of his weapons.

That Walter entered with full ardor into the spirit of the Noviceship is clear from the testimony of his director and Superior and from his later life. He was frequently commended and held up as a model to others.

> Father Drum was a Novice when I was a Junior [writes another Jesuit]. I am sorry to say I never was close to him or shared his confidences. But I admired, as we all did, his tremendous zeal and earnestness. He entered the Society at full speed and never let up. He has a high place in Heaven.

However, "the annals of peace are brief" and there is little to record during this period. In the Society's course of training, every process has been scientifically worked out, and the raw material passes through various stages of preparation, of sifting, of grinding and refining to come out a finished product —white as the wheat to which Ignatius the Martyr wished himself to be ground.

Only a detail here and there in the process will strike the eye. In his second year, Walter Drum was chosen, because of his maturer age and judgment, to the office of Beadle, or Manuductor as it is called in the Novitiate; and in that capacity had to assign the various daily offices of the Novices, to ring the bell for exercises, and to be in general the vehicle for communications or instructions from Superiors. His fellow Novices still remember the deep bass voice that now and then boomed through the stillness of the ascetory; and a smile would on occasion follow the resonant *Hodie admonendum est*, or, *Ambulatio generalis*, or, *Manualia hoc mane*.

He had learned Spanish and a little Italian during his student days at Las Vegas, and his notes of these Noviceship days show that he did not neglect language studies even at this time. His spiritual reading—of Rodriguez, and other books—was done in the original Spanish, and, as was his custom then and always, he made many excerpts from his reading and wrote out long passages in Spanish, grouping them under special headings.

When one subjects himself to an honest scrutiny, or is willing to submit to the candid judgment of others, there are many surprising discoveries possible. As in the case of every one else who has gone through the experience, Walter Drum found that even his first attempts at self-scrutiny revealed unexpected peculiarities. He made, therefore, valiant attempts, but in the end, vain ones to change or modify his "million-dollar walk," and his declamatory manner of conversation, and also his warmth in discussion and argument. But the sense of humor did not fail him, as we may see from some of these reminiscences of his own:

> We have an exercise in the Novitiate which we call *lapidatio* (stoning). The Novices sit side by side in two rows and the victim kneels between them. The first time I was down on my benders for the stoning, the Novices told my defects, threw little pebbles at me. Then the Master of Novices dropped a boulder on me. He made me feel like a penny that a railroad train has gone over. He said: "Well, I think he is rather vain in the way he parts his hair. He does not part it like an ordinary Christian on the left." The next time I had a stoning, I was ready with a prison crop, as close as could be. In all good will, I decided not to be vain, and thought the best thing to do was to get rid of the thing. But the boulder came as usual. The Master of Novices finished up the performance by saying: "A Novice should remember that he has not ceased to be a gentleman by becoming a Jesuit. It is an insult to his fellow Novices to appear as if the Novitiate were a penitentiary. He is unattractive enough without the help of the axe." Well, then, after all good will had failed, I went to him and said, "Father Master, what shall I do?" "Part it on the left." "Father, it won't stay; there's a cowlick, it won't stay on the left." He said: "Part it on the left." So, for three years it was topsy-turvy. He always took that line with me, and finally I realized what he was doing. In the beginning, I had no idea.

> He made me lampideer—I had to fill the lamps with oil. Bring the oil from about fifty yards away, in February. The result was that my hands were chapped and dirty. I simply could not keep them clean. I knew nothing about filling lamps, and when it came to filling fifty or sixty lamps, I was a hopeless mess. So he always used to take hold of my hand, and fling it away in disgust, and say: "Oh, do be clean," until finally I realized what was going on and then it did not disturb my equanimity. When we realize that the stoning is stoning, and meant only for our good, then we do not mind.

Walter Drum pronounced his First Vows in the Society on the Feast of the Nativity of the Blessed Virgin, September 8, 1892, and with a group of others crossed over to the Juniorate to begin a course of classical studies. The principal aim of this course, which usually takes two years, is to prepare the Scholastics for the teaching of Latin and Greek and other subjects, in the colleges and high schools of the Society. In the case of Scholastics who are older than the average or who have already completed a college course, one year is often considered sufficient for the purpose. So it was decided for Mr. Drum, and, after one year of Juniorate, he was sent to Woodstock for the study of philosophy.

We find a list of graces to be prayed for, written down at intervals, from the days of the Novitiate up to the time of his Tertianship; it is a revelation not merely of the yearnings of a young recruit eager to go forth to the battle of the King, but we also see something of the loyalty and affection that animated his youthful heart for the Society, for his family, and those near and dear to him; while the very greatness of his sincerity is manifested by the generosity of his hopes and aspirations and the favors he prayed for.

> A. M. D. G.
>
> May the Sacred Heart of Jesus be everywhere loved.
>
> Sweet Heart of Jesus, burning with love for me, filled with graces that You will pour forth upon me, if I only ask and receive them, grant I beseech You, these favors that Your humble servant and loving follower begs of You, accompanied by his spotless Mother, the Angels, St. Francis Xavier, Sts. Francis Borgia, Regis, Hieronymus, de Sales, and Assisi, Sts. Stanislaus, Aloysius and John, St. Walter, Sts. Gertrude and Catherine and all the other

Saints of the Society and members of the whole celestial court, my protectors. Amen.

Graces

To die in the Society—St. Ignatius.

To make my last Vows, and to say Mass till my death —St. Francis Xavier.

To live in the true spirit of the Society—Guardian Angel.

To receive from God the vocation and the grace to go on missions outside the Province—St. Francis Xavier.

That this Province especially may increase in number and spirit—Our Lady.

That the Spiritual Exercises of our holy Founder may be studied well by Ours, and very often given with great fruit to externs—St. Ignatius.

That I may know the King more intimately, love Him more ardently, and follow Him more closely—Our Lady.

That I may have a burning, desiring love of the Sacred Heart of Jesus—Blessed Margaret Mary, Fr. de la Colombière.

That my parents and the family may all live, if it be the Holy Will of God, to see me ordained a priest of the Society and to hear my first Mass—Sacred Heart of Jesus.

That I may always have an ardent love for Jesus in the Blessed Sacrament—St. Francis Borgia.

That I may pursue my studies with great zeal, merely for the greater glory of God—St. John Berchmans.

That I may have a keener perception and greater love of holy purity day by day—Our Lady, St. Aloysius.

That I may have a great love for the *agendo contra*—St. Francis Jerome.

That I may increase daily in the strict observance of all our rules and customs—St. John Berchmans.

That I may ever practise fraternal charity, cheerfulness, humility, and mortification—St. Alphonsus Rodriguez.

That I may always have a great love for penances and practise them—St. Francis Borgia.

That I may be true to all inspirations of the Holy Ghost, and cultivate a great love for Him—Holy Ghost.

That our Novices may increase in spirit and number—St. Joseph, St. Stanislaus.

That God may guide our Father General, Provincial, and Rector and all other Superiors to rule in all things *ad majorem Dei gloriam*—St. Ignatius.

That the Catholic Church may prosper against all her enemies—St. Gregory VII.

That the missions of the Society may increase in prosperity, religious zeal, and numbers—St. Francis Xavier.

That God may grant eternal happiness to all those who die in the Society—Our Lady.

That Jesus may grant the canonization of Blessed Peter Favre and Blessed Peter Canisius and of Blessed Margaret Mary and of Father de la Colombière—Sacred Heart.

That I may be always devoted in a great way to my patron saints for life and for temporary periods—St. Stanislaus.

That all of my patrons may ever be watching over me until I join them in heaven—St. John Berchmans.

That I may ever have a great love for Jesus on Calvary—St. Mary Magdalene.

That my Father, Mother, Brothers and Sisters may all meet me in heaven—Our Lady.

That all heretics and infidels may be converted—St. Paul.

That I may have a great devotion to the Guardian Angel placed over me and those placed over others—Bl. Peter Favre.

That sweet concord may reign in our family—St. Joseph.

That God may grant a happy death to all the members of the family—St. Joseph.

That the family may advance rapidly spiritually—St. Joseph.

That Jesus may take care of the vocation of each and every one of them—Sacred Heart.

That I may acquire a zeal for my own perfection—Jesus Crucified.

That I may be very zealous for the perfection of the souls of others—The King.

That all the members of the family may receive all the spiritual and temporal favors they are in need of—St. Joseph.

That God may take care of the families of my relatives—St. Joseph.

That I may never deliberately break the least rule of our Society—St. John Berchmans.

That Jesus may send me many humiliations and great opprobrium and dishonor, and the grace to bear up against it, and to suffer joyfully with Christ suffering—Sacred Heart of Jesus.

That I may always have great devotion for the Saints of God—St. Stanislaus.

That God may grant good health to all the members of the family—St. Joseph.

That Christ may be the more honored for the sufferings that He underwent—Sacred Heart.

That I may be gifted with the virtue of recollection—St. Stanislaus.

That I may die without deliberately breaking a rule from this time on—St. John Berchmans.

That I may have tact to lead others into conversation on spiritual matters—St. John Berchmans, Bl. Peter Favre.

That I may receive from my King a very great knowledge of the Spritual Exercises, and thus be able the better to spend myself for the promotion of the greater glory of God by the salvation of souls—Our Holy Father.

That our Novices may always make the Great Retreat with much fervor and fruit—St. Ignatius.

That I may acquire a more perfect imitation of my Jesus *mitis et humilis corde*—St. Francis de Sales.

That the King may grant me the grace of vanquishing the least temptation to vain glory—Our Lady.

To live cheerfully and according to a fixed method, serving our King with a constant expansion of heart—St. John Berchmans.

If it be allowed by the Infinite Mercy and Love of God, that I may live a life of perfect observance of my rules until my Tertianship, that I may then make the vow of perfection which the Ven. Claude de la Colombière made; viz., never to violate a rule—Sacred Heart of Jesus, Our Lady, St. Ignatius, St. Francis Xavier, St. John Berchmans, Ven. C. de la Colombière.

That I may observe my Vows most scrupulously and may advance very much and continuously in the solid virtues of Poverty, and Chastity and Obedience—Sacred Heart, Our Lady, St. Ignatius.

To be filially devoted to Mary, my Mother, and to advance rapidly in perfection on account of this devotion—Our Lady, Sacred Heart, St. Joseph, St. Stanislaus, St. Aloysius, St. John.

To grow strong and have good bodily strength in order that I may work with much energy all my life for the salvation of souls and die generously fighting the battles of the King—in harness—St. Joseph, St. Raphael, St. Francis Jerome.

That I and my Brothers in the Society may be tenderly devoted to the Sacred Passion of Our Lord—St. Francis Regis, St. Catherine of Sienna.

For family,—to be excellent Catholics, to go to Communion at least once a month, to live in peace and love with each other, to live and die in the state of grace, to enjoy eternal glory with me, to die fortified by the last Sacraments. [Here follows mention of each of the family by name, and the special graces he prays for for each.]

It would appear that he was permitted, towards the end of the Tertianship, to make the Vow of Perfection, mentioned in this list, never to violate a rule, in imitation of Ven. Claude de la Colombière.

There are among his notes several sermons that were written and preached in these early days, one on St. Stanislaus which he preached in his second year, on the eve of the feast of the Saint, November 13, and another for the feast of St. Aloysius the following June 21. In the Juniorate, one of his sermons is on the Blessed Sacrament; another on Faith: we should always be on our guard lest the gift of faith be given to us in vain. As we read them, we see even at this date the mainsprings of his whole future life, and the genesis of that spirit of loyalty to the Faith and to the Society that was to be his most conspicuous characteristic.

CHAPTER V

WOODSTOCK AND PHILOSOPHY

From all accounts, the decade of 1890 was an interesting age at Woodstock. It was the Woodstock of Father Frisbee and the famous "W. W. C.," the "Woodstock Walking Club," whose sturdy feet cleared the undergrowth and kept open the scores of picturesque by-paths in and out of the woods of Maryland. The long walks on holidays kept the men in prime condition, and the personality of Father Frisbee was perhaps the main influence that spread a genial spirit of comradeship and good-fellowship through generations of Woodstock Jesuits. This, too, was still the period of Father Sabetti and the busy landscape gardening of the grounds of Woodstock, the results of whose labors was to produce what was once called by a Baltimore paper "one of the attractive parks in the environs of our city." Father Edward V. Boursaud, S.J., was then Rector and welcomed the party of young Scholastics from Frederick in September, 1893. But late in the Fall of the same year he was succeeded in that office by Father Joseph M. Jerge, S.J.

Mr. Drum was at this period frequently disturbed by racking headaches, due in part to the same weakness of stomach that had afflicted him in the Noviceship. But it was also superinduced by a rigorous fidelity to fasting and other penetential exercises which he was permitted to practise—to excess. as he afterwards admitted, for he nearly undermined his health. But, by dint of strong will-power and intense application, he succeeded in doing exceptionally well in his studies, and twice appeared for a "Defense" at the general Disputations held before the whole seminary, once in psychology in February, 1895, and again in cosmology in the following November. He showed his willingness to enter into all plans for the general entertainment, and for two successive years took charge of the Christmas arrangements; and, as he was a good vocal soloist, he took a part in the activities of an ambitious glee club which once presented a modified form of Gilbert and

Sullivan's cantata, "Trial by Jury," much to the delectation of the community. His earlier experiences and his companionship with his father and the family in their travels had given him a large store of anecdotes and a ready facility for story-telling; and his memory had also gathered up a varied assortment of negro songs and melodies. This versatility was coupled with an odd humorousness that was all his own. Besides, even thus early he showed his talent for preaching, in the customary round of sermons which, according to immemorial traditions, on certain days of the week, take the place of reading in the Refectory. He entered actively into outings and picnics on vacation days. One of his close friends records how Mr. Drum used to expostulate with him for not joining more freely in such excursions. "You are willing enough to help anyone who comes to you,—in matters intellectual, in studies and the like, but you do not help others to enjoy themselves." Perhaps nothing reveals in better light the man's genuine zeal and charity than this generous desire to help on the spirit of good-fellowship in the community.

At one time or other, Walter must have torn up and thrown away a great many of his notes of these early days, but there remain a few scattered pages, torn from different note books, and kept for some reason of his own, probably in order to use the thoughts and develop them at greater length for sermons or retreats. They are of interest now chiefly as revealing the aspirations that filled his mind in the first years of Religious life, and to a certain extent, the vigorous spirituality of the character that was being gradually developed in him. The following are a few random selections:

> The Epiphany. The Magi brought their gifts from afar, for they knew not precisely where they would come upon the Child. Thus, begin young, afar off, to gather in your gold, frankincense and myrrh against the coming of Jesus, when He shall take your soul to Himself. For, who knows when He will come, and when you will have to give up the few gifts you have gathered?
>
> St. John Chrysostom. His zeal in defending and preaching and spreading the Church. We Jesuits who are so bound to the Holy Father ought to give ourselves most zealously to the Holy Church's works. She needs good preachers. The gift of eloquence is not enough. With-

out it, one can, by dint of personal sanctity, accomplish what others cannot with it. St. Ignatius brought tears to the eyes of even those who did not understand him.

St. Joseph. He was filled with all virtues and thus made a good husband for Our Blessed Lady, and a good Foster Father to Jesus. How he loved them! What a supernatural tone this love gave to his life. Mark what a care he had for Our Lady's fair name. How do I act in this regard? Does not she want to see me pure and filled with love for her virtues, because of her name that I have taken to myself?

St. John at Latin Gate. Grace to come out from all trials and temptations more strong and spiritual. *Hic est ille discipulus quem diligebat Jesus.* What was it to love Jesus? What resulted from the love of Jesus? The wished-for prize of martyrdom did not come. But years of hard work and years of comforting of souls was the life work of St. John. He was so near to the Sacred Heart. Then grow like unto the Sacred Heart in your every aspiration and thought.

His consoling thoughts, the joys of the past—reclining on Jesus' bosom. The joys of the future—The eleven will be seen! and the Lamb! and Our Lady! Sighs and aspirations for the Beatific Vision. Prepare for Heaven! All for Jesus!

St. Ignatius. He was a giant in his great love for Our Lord, Our Lady and the Society. His personal love for Our Lord was what filled his soul with strength. This love brought on a great love for the Society, his little body of *insignes* who were fighting as companions of Jesus, to propagate the faith of Jesus, and to be the chief support of the Church of Jesus. Why did he so love the Society? Because she was, next to Christ's Church, the most cherished means, to him, of engendering in the hearts of men this same personal love of the King. Therefore, does he today pray with a most instant prayer that Jesus give this pure love to his dear ones on earth, that these dear sons be true to their call. O holy Father, grant me this love of Jesus. Ask it for me from Jesus! May my heart beat and every nerve twang this one sweet note: love for Jesus! May love fill my whole heart and bring forth its fruits in my soul. May these fruits be the pearls that should bedeck a true son of Loyola.

St. Francis Xavier: Dying on the island of Sanchan, under an open shed. Grace to see and understand his love, and the strength to carry it out in my own life. St. F's. love for the Society shown: (1) because he gave up the world for it; the struggle with vain glory; the lasting conquest. (2) He gave his labors and spirit of zeal for it,—incessant mortification, earnestness. (3) He gave his life, died without even his Jesuit companions; (consider his love for St. Ignatius and for Companions); and without Viaticum. His love for Our Lord,—its entirety, generosity, constancy, humility. Rewards. Pray for the vocation to foreign missions and for the *Votum Perfectionis.*

Sacred Heart. Grace to live in the spirit of the Third Prelude of the Second Week. Devotion to the Sacred Heart means this knowledge, love and service of Our Lord. Know Him in His inspirations, in study, prayer, work, etc. Do not know yourself here, but recognize Him. Love Him here with a self-sacrificing, an earnest love. Why be devoted? That is our vocation. Our name: Companions of Jesus—a brotherly love for Him. Decree of the Congregation, consecration of the Society. Fruits of this devotion—all will be love and joy. My soul will receive its humiliations, I shall be despised; but, sweet Jesus will be mine forever! What more do I want? How sweet to possess Jesus!

Years afterwards, during the brief visit that Father Drum paid to Ireland, he met an old Irish lady, a cousin of his mother, who gave him a little versified form of this Third Prelude, which took his fancy so much that he copied the lines and used them frequently in later retreats. Many have thought the verses to be his own translation, but they have been used by so many Retreat Masters that it is now difficult to discover the author. The verses are as follows:

Sweet Heart of Jesus,
Grant us, we pray:
To know Thee more clearly,
To love Thee more dearly,
To serve Thee more nearly,
Just for today!

St. John Berchmans. By his love for solid virtues he perfects himself in the religious life. Studies must now replace the little virtues of the Novitiate. These sweets are over,—at least, the means are not now so favorable. So let your heart leap towards the Blessed Sacrament whenever the bell calls you from study.

Be faithful to Jesus Crucified. Such is the spirit of the Society. And vile it were for this dull spirit of mine not to be moved by a love of imitation, by this great suffering in my Jesus. How often have I the chance! What humiliations I shirk! and yet, I have a special vow! By a life of cheerful and humble exterior deportment I will help the interior humility. How am I as regards studies? Am I not filled with pride in all ways?

Be faithful to Our Lady. What love she merits from all and from me. I am consecrated to her service, and yet how negligent in the service which is most dear to her. Love of Jesus Crucified and of prayer!

Be faithful to the Society. What a noble, self-sacrificing body of heroes I belong to! How do I treat them?

St. Francis Borgia. *Franciscus peccator*, and he meant it too! What a chance for all the world loves! Would I have given up all this? How do I look upon the world now? Do I love humiliation? Do I tell my troubles to others and seek human consolations? Do I live united with my humbled Jesus? Chastity,—his life in the world, and in the Society, angelic. Charity,—how did he speak of others and treat with them? His loving heart saw their virtues and passed by all their faults. Whereas I,—the contrary! !

St. Catherine. Truly wise. *Timor Domini initium sapientiae.* I strive vigorously merely after worldly wisdom. Study hard, especially philosophy,—pray much that you may become very learned, may master the matter and grasp it firmly and retain it,—but all for God's glory and the Society's work and end. Pray most instantly to St. Catherine for these favors, and to be kept in an humble position in class, to be ridiculed and laughed at therein, never to be appointed for Disputations, to fail always in recitations, to be considered ignorant,—a fool. Pray much about your vow never to avoid a humiliation deliberately. Pray for the entire class, especially for those that are holy and study hard, yet do not grasp the matter, —pray for them by name.

Virgin—whole Society. My vow. Pray much for angelical purity and love of Mary.

Martyr—Missions and martyrdom. Vow of perfection,—unbloody martyrdom.

Souls in Purgatory. How will the light shine upon them? They must be happy in anticipation, but most sad in long expectation. Their works will follow them. I ought to keep this well in my mind. My works are, one and all, good and bad, to follow me, yes, follow me far beyond the grave. No hyprocrisy then. O my Jesus, all my sins are known to You and me. You hate them. Teach me, too, so to do. May I now begin to bring souls to Thee in every way. Here is one way to repair the past,—free a poor soul. What a joy to give God this soul!

Sermon on Mount. We all dream dreams at one time or another of our life,—day-dreams, night-dreams, play-dreams, and pipe-dreams; and the things we dream of sometimes come to be a very part of all we are. The school boy wishes he had been with Caesar, in the conquest of all Gaul; or with Alexander, in the unifying of all Greece; or with Napoleon, in the turning of the Russian flank at Austerlitz. All these dream-wishes pass away; Alexander, Caesar, Napoleon,—become names and no more. Such is not the case of every dream-wish. In boyhood days I had the flitting day-dreams; the mere memory of them is all but gone. Later, I had a staying day-dream; I have it now; in all the strength of boyish fancy-flight it comes to me,—I wish I could have seen the dear, dear Lord, as He stood on the hill-top, outside Capharnum; I wish I could have sat with John at the very feet of Jesus; have gazed up into those deep, dark, loving eyes; have watched the earnest face, full of glow and enthusiasm; heard those words of the Sermon on the Mount, those words of tidings new which changed the whole world's view-point,—Blessed are the poor in spirit.

CHAPTER VI

THE TEACHING PERIOD

After the completion of the third year of philosophy, there usually follows in the Society, a period of teaching, the Regency, as it is more commonly called, in one or other of the colleges or high schools of the Province. The young man who goes through the preliminary training of the Juniorate and of the years of philosophy is sufficiently equipped to take his place as a professor, granting of course a reasonable time for more immediate preparation, in any of the classes of the high school, and often also in the Freshman and Sophomore classes of the college.

Under present arrangements there is provided a regular summer-school course for the Scholastics who are still in the Regency, during which the work for the coming year may be outlined, while at the same time certain of the more experienced teachers are assigned to conduct the courses of the summer school and help the new recruits in whatever prospective difficulties may occur.

But in the days when Mr. Drum found himself assigned to teach the first-year high-school class at St. Francis Xavier's College, New York, all the Scholastics had to make their own proximate preparations, unaided, for the work of the classroom. True, the previous training is entirely along the lines of the "Ratio Studiorum," which is the traditional basis of the Jesuit educational system, and their own researches could familiarize them with its methods, at least in practice; but then it is no easy task for an inexperienced youth to prepare, unaided and of his own initiative, an outline of class-work which he is to follow for a whole year, even with the program of studies in the college catalogue for his guide. Nowadays, the task is happily simplified, by the aids that are available in the summer school.

Mr. Drum found his teaching at St. Francis Xavier's very congenial. There is a brief reference to this first year's work in the notes of one of his retreats.

> In my first year of teaching everything went along fine. The boys were industrious, there was a great deal of study and recitation and I certainly looked forward to the happiest results. I liked the boys and they liked me. We got along splendidly together. Indeed I taught with such vigor that it would have brought on nervous prostration if it had gone on. Three of the boys are now Jesuits; two of them are at Woodstock now, one of them has just begun his Regency. Everything was such as to make me quite satisfied with myself.

But he speaks in an entirely different strain of his experience with the class of Second Year High at Georgetown College, to which he was transferred for the following year, 1897-1898.

> That year [he says] everything went wrong. My class, they told me, was exceptionally poor. I had millionaires' sons and baseball players among them, and but few who cared to study. But that year did me personally a great deal of good. It was my triumph of failure. The work was very hard with that particular class, —five hours a day, and the rest of the time prefecting the corridor. Day after day the struggle went on. Discouraged and weary, I was on the point of reporting all to the Superiors. But the thought of suffering in silence for love of Our Lord held me a little. I used to go up to my room, after the day's teaching, kneel down and for fifteen minutes make a better meditation than I ever made in the Novitiate. I used to take up my Crucifix and say, "Dear Lord, everything is as it should be. It is now my turn to be on the Cross. You will never again be on that Cross, so that is enough for me. I am grateful to be able to take Your place even in this little way." That year meant a great deal to me.

Early in July of 1898, after Mr. Drum's trying year at Georgetown College, there came stirring news from the war front, in Cuba. It was proud news for the nation, but fraught with grief, as war ever will be, for the families of the fallen. Long after this time, in one of the sermons of the *Tre Ore* given in the Immaculate Conception Church, Boston, Father Drum describes his reminiscences of the days when the reports were coming from Cuba:

> One of those who died in the battle of Santiago was a Catholic officer whose heart was true to his God as it was true to his country. He died in the advance of his

troops. His family learned of his death, and grieved,—grieved that he had passed away, grieved that they had not heard his dying words nor cheered his dying moments. Some ten days after the battle, the family received a letter,—their dead father's last letter,—his last message written a few moments before the fight was on and found on his person after death. In his last words, he told his dear ones that he had just been to Confession, that he went to battle ready to die. He wrote the name of each, and bade them stick together, and called upon God to bless them. How the family now cherishes that letter, how it clings to the memory of the last words of the dear father dead!

The Catholic officer was his own father. The gallant Captain Drum was the only officer of his brigade to fall in the terrible engagement of July 1. He made the supreme sacrifice at El Caney in the advance that followed the battle of San Juan Hill, some three hundred yards further from that locality in the direction of Santiago. While standing to give the word of command, he was shot through the breast, and lived but a few moments afterwards. He was buried in a temporary grave at the foot of San Juan Hill, and in the following month, the body was disinterred and conveyed to Montauk and thence to Boston. It now rests in Arlington Cemetery, Washington, D. C.

Among the many expressions of honor and condolences received by the family of the distinguished soldier, there was a letter from the Adjutant-General of the United States, on behalf of President McKinley, who, in accordance with official courtesy, presented to Mrs. Drum his own and the nation's condolences in her noble sorrow, and offered a commission to any one of her sons who should choose a career in the Army or Navy. All the brothers after long consideration, waived their claims in behalf of Hugh, the youngest son, who was not yet nineteen years of age, and had just completed the Freshman year at Boston College. This decision was finally made in deference to their father's known wishes that Hugh should follow the military career.

A preliminary examination was necessary and accordingly Hugh Drum spent the rest of the summer vacation in hard study, under the coaching of his Jesuit brother, and, at the end of August, succeeded in passing a splendid examination. The question of his extreme youth was brought up, but was waived by special consent of the President, and Hugh received his commission as Second Lieutenant, and was appointed to the Twelfth U. S. Infantry, on September 9.*

---

*"The next to youngest officer in the United States Army," Hugh A. Drum was able to find himself in immediate service and to manifest his ability as a leader of men. In February of 1899 he went with his regiment to the Philippine Islands and remained there until April, 1901. During this time he took part in the various campaigns and skirmishes under General MacArthur. He returned to the U. S. for a few months, and in January, 1902, again went to the Philippines and entered on the Lake Lanao campaign under the then Colonel F. D. Baldwin. He had been promoted to First Lieutenant and was later appointed Battalion Adjutant, and at the conclusion of this campaign was recommended for brevet as Captain and appointed Aid-de-Camp on the staff of General Baldwin. He also served in a third Philippine campaign in 1908. By dint of constant study and strict attention to every detail of military duty, he merited high commendation and surpassed every obstacle to his progress. He graduated with honor from the School of the Line, and in 1911 was graduated from the Staff College, 1912. From 1913 to 1914 he was Assistant Chief of Staff to Generals Carter and Funston on the Mexican border and at Vera Cruz. From 1914 to 1916, he was instructor in military art at the Staff College, Leavenworth. Then he became Assistant Chief of Staff to General Funston, on the Mexican border, and on the death of the latter, was retained in that capacity by the then Major General Pershing, who succeeded General Funston. He was promoted a Major in May, 1917, and was appointed to the General Staff under General Pershing. Major Drum was the first General Staff officer of the American Army to arrive in France; upon him devolved the duty of choosing the ports of entry and the places of training the American troops, and later, as Chief of Staff of the First Army, he had charge of organization and operations in the field. It was he who planned and directed the execution of the St. Mihiel and the Argonne-Meuse battles. He was shortly promoted to the rank of Lieutenant Colonel, and at the close of the war held the rank of Brigadier General N. A.

General Drum was thirty-nine years old at the time of his return to the United States, and was known as the youngest Chief of Staff of any Allied Army. His decorations number the D. S. M. from his own country—for which he received the citation from General Pershing—the Order of the Crown of Belgium, the French Commander of the Legion of Honor and the Croix de Guerre, and the Italian Commander of the Crown.

But, most of all, his friends treasure a direct citation from Marshal Foch, given to only a few officers in the Allied Armies, which reads:

"On approbation of the General Commander in Chief of the A. E. F. in

But in the next September, in the beginning of the school year, 1898-1899, Mr. Drum was sent again to St. Francis Xavier's. During this and the following year, he taught the class of Fourth High.

At St. Francis Xavier's, Mr. Drum's dramatic ability was brought into conspicuous play. It has always been considered a distinct feature of the Jesuit system of studies to develop in their students the power of pleasing and convincing expression, by means of practice in elocution and in dramatics. In both these fields Mr. Drum achieved truly remarkable success. The college had some years before gained a national reputation by its presentation of the famous Latin play, *Duo Captivi,* of Plautus, which merited a medal of recognition at the Columbia Exposition, in Chicago. Many declared that the prestige of Xavier was more than upheld by the imposing dramas which were enacted in the college theater, under Mr. Drum's direction. He was "Moderator of the Dramatic Society" during the three years that he taught classes at St. Francis Xavier's, and the plays which the students then produced were honored

---

France, the French Field Marshal in Supreme Command of all the Allied Armies, is mentioning in despatches:

"Drum, Hugh A., Brigadier General, Chief of the Staff, First Army:

"Being called upon to act as Chief of Staff, First Army, at the time the latter was constituted and having been entrusted at the same time with the study of a series of major operations in the St. Mihiel region and between the Argonne and the Meuse, he displayed exceptional qualities of understanding and sound judgment, as well as unflinching energy. He shared greatly in the success of the above operations.

Allied General Headquarters, the 4th of June, 1919
The French Field Marshal in Supreme Command of the Allied Armies.
Signed: Foch.

On returning to the United States, there were general demotions following the muster from service of the National Army and the rearrangement of the Regular Army, and General Drum was first made Assistant-Commandant and later was given full charge, as Commandant, of the General Service School at Fort Leavenworth, with the rank of Colonel. But in the Fall of 1922 for distinguished services he was made a Brigadier General. In the following February he was appointed to the command of the military defenses of New York City, and in the early summer of 1923 organized a series of elaborate military maneuvres to test the defenses of the metropolis that won wide praise and admiration.

with most laudatory reviews in the secular press, newspapers like the New York *World* and the New York *Evening News* declaring these plays to be far and away above the standard of ordinary amateur dramatics: while the most unstinted praise was given them by Catholic journals, like the *Irish American*, the *Catholic News*, the *Freeman's Journal*, and the Boston *Pilot*.

One of the students who took part in the production of Julius Caesar, and who is now a priest in the New York Archdiocese, tells of watching Mr. Drum during the rehearsals for this and other plays:

> Mr. Drum would speak and act every part himself to show us what had to be done. And he never used a manuscript. He had every line of the whole play memorized, and in portraying each part he seemed able to live it. He was slight of figure and seemed a frail man, but he had a powerful voice. We used to see him sometimes, between classes, walking up and down in the open air, with his majestic walk, chest up, thumping his chest and taking deep breathing exercises. Did not mind in the least who was looking on. We took to imitating him ourselves. He was the type for preaching, lecturing. I was surprised that he was ever selected for Scripture work, which seems to me to require a book-worm type of man.

Another of his students, now a Jesuit, recalls that Mr. Drum told him on one occasion, that he had spent altogether about thirty hours in the Astor Library, studying costumes for one of his plays.

We may form some idea of the wide recognition that these plays received, from the following citations from an article which appeared in *Werner's Magazine*, for August, 1900, entitled, "The Catholic College Play," by Ruth Everett, who used all her illustrations and examples from the plays presented at St. Xavier's College. After explaining the wide difference in the character of the productions put forth by non-Catholic as compared with Catholic Colleges, the author goes on to say:

> In the Catholic College, our old friend, the dramatic coach, rises to the dignity of, and is known by the rather high-sounding and ponderous title of "Moderator of the Dramatic Society." There is, however, no more difficult position in the entire work of the college.
>
> The present moderator of the Dramatic Society of St.

Francis Xavier's College is Mr. Walter M. Drum, S.J., a son of the late Captain John Drum, who was killed in battle beyond San Juan Hill. What Mr. Drum does in St. Francis Xavier's College is done by every other man who holds a similar position in any Catholic college. Two plays are given during the year, one by the boys in the preparatory department of the College, and one by the collegians. The first play is produced about the time of the Xmas holidays, the second at Easter.

Months before the time, Mr. Drum selects the play, which he must adapt to the requirements of the Catholic college. For here, the female parts have to be eliminated, and the effect that the female characters have on the action of the play, must be conveyed to the audience by incorporating their story in the lines of other characters of the original play. Sometimes this is not possible, and an entirely new part has to be written in, thus necessitating many changes throughout the entire play.

While in the Catholic college a reasonable laughing farce, like "Box and Cox," is sometimes given, this is a departure from the rule and the well-defined purpose of the college, and is not encouraged. Burlesque or light opera would not be tolerated. The professors say that if the boys want to do anything of this kind, they must find other places than the college theater for their show. So, in all seriousness, the legitimate drama is adhered to, education and liberal culture being the aim kept steadily in view.

The last play given at St. Francis Xavier's College was King Henry V, in which thirty-six college men had speaking parts; and before the battle of Agincourt, the stage,—which is the finest of any non-professional stage in the city of New York,—was filled with soldiers. Shakespeare's play has about 3300 lines which were cut down to half.

Before memorizing the lines, and before a single rehearsal, every boy in the play must make a careful study of the customs, dress, and manners of France and England at the time of Henry V. The boy who is cast for the king must write an essay on the King Henry of Shakespeare and the King Henry of English History. Nothing must be overlooked. The Dauphin of France, in derision, sends King Henry some tennis balls. In the boy's essay he is expected to explain the origin of the game of tennis, and how the modern game differs from that of the days of Henry V. If Mansfield or any other celebrated actor were playing King Henry V in New York previous to production at the college theater, the dramatic society would be advised to go and see the play and take careful notes.

The mid-winter play this year was "The Bells." Several circumstances conspired to make this a most finished amateur performance. First of all, "The Bells" is a one-part play; no one in the cast is any more than a feeder to the star. Next, immediately before the play was rehearsed, Sir Henry Irving came to produce it in New York. Mr. Drum went with the Dramatic Association to witness the performance. Lastly, but most important of all, the man who took the part of "Mathias" was Thomas A. Brennan who is gifted with exceptional emotional abilities.

Last year one of the plays was King John, in which forty-one boys had speaking parts. The King John of the play wrote a very good essay portraying the King John of Shakepeare and the King John of history, as also the quarrel, and its causes, between him and Pope Innocent.

Two years ago Bulwer Lytton's "Richelieu" was played. Imagine, if you can, the supreme courage of the amateur who is forced to pronounce the Cardinal's great curse, and hypnotize his audience into the belief that Julie is being drawn "within the awful circle of our solemn church," for, mind you, no Julie is there at all. Why it would have driven Edwin Booth off the stage with nervousness. But the Catholic college boy gets through with it and seems to like the effort.

Another point about which the Catholic college is extremely careful is that of costumes, which must be historically accurate. If ever a slight deviation is made, as was done in the "Comedy of Errors," the matter is explained to the boys. Shakespeare's "Comedy of Errors" is so full of anachronisms that it has given rise to considerable discussion as to whether the costumes should be those of the Byzantine Christian period or the Italian style of the Elizabethan period. It had been produced on a most elaborate scale by many great actors,—Augustin Daly, Robson, and Crane among them. At the time it was decided to give this play in the college, Mr. Drum consulted both with Mr. Robson and Mr. Daly. The former advised the Elizabethan style of costume principally because the boys, who averaged sixteen years of age, would look very graceful in the dress of the Byzantine Christian period. Mr. Daly, however, favored the other style. As there was good authority for either period, it was decided to choose in favor of grace and beauty, and the Italian dress of the Elizabethan period was used.

One of the students who took part in the productions, and is now Headmaster of a distinguished New England School writes the following letter of reminiscence and appreciation:

I first met Father Drum, who was then a Scholastic known as Mr. Drum, at the beginning of my freshman year, September, 1896. He taught me Mathematics—God save the mark—or I should say, labored at teaching me Mathematics. It was something about cube root, and one problem would cover a whole blackboard, and there was nothing but m^3, n^3, x^3, *and* y^3. I think he was almost as poor a teacher of Mathematics as I was a student of that recondite subject, but he certainly was a hard worker.

I remember his splendid, round voice, his flashing black eyes, and his eager, nervous, crouching attitude as he hurled those cubes at our heads. I imagine his superiors must have given him the job of teaching Mathematics for the good of his soul, for even though I was a student under him in this subject, my most vivid impression of him in that year is not so much as a teacher of Mathematics, but as a teacher of Greek in the first academic class across the hall from where we studied.

It was positively thrilling to walk along that corridor while he was holding a Greek class. I knew something of contract verbs and *mi* verbs myself so that I was in a position to listen with wonder and awe and appreciation to the magnificent intonation he gave to the paradigms for the benefit of the boys who studied under him. He had a most oratorical way of singing the contracted vowels, but what astonished me most was that he used to pronounce the *mi* verbs according to accent, whereas good old Father Keveney, who had nursed me through the early stages of Greek, used to hit them hard on the long vowel. It would not have mattered to me what he was saying in Greek so long as he said it so impressively and well. You will understand from this that even that year my most vivid recollections of him are not personal or physical, but the recollections of a wonderful voice intoning wonderful sounds through an open transom in a dark corridor.

I think there was some added glamor by reason of the fact that he was known to be the son of Captain John Drum, Commandant of our battalion, who, college tradition said, was known among the Indians of the far west as "Thunder Voice." Naturally, we considered Mr. Walter Drum the worthy son of a worthy father.

At that time, however, I had no part in any play, but

I won the Elocution prize at the end of my sophomore year and this stroke of luck must have brought me to his attention for in the following autumn, when I was a junior, I remember distinctly that I was called out of class into the corridor and he was waiting on the other side of the door. He told me that he had selected me to play Falconbridge in King John. I must have asked him some such question as Flute, the bellows mender got off—"What is Falconbridge, a wandering knight?" For he gave me to understand that it was a mighty hard job for a young man to do, and I went back to class with trembling knees and a hollow stomach. Naturally, during the rehearsals that followed, I got to know Mr. Drum quite well.

I have since produced plays myself and really appreciate the stupendous labors he performed in getting out this production. He went into it, every ounce of him, with complete abandon. He acted every fellow's part for him time and time again, and he never seemed to wear out, though he was working at his classes and doing all the mechanical preparation for a rather heavy scenic production in addition to the hours he spent in rehearsing us in the college theater. Towards the end of the rehearsals he secured the services of that charming old gentleman, Mr. Fred Williams, the stage manager of the Lyceum Theater, then on Fourth Avenue, and we had several final rehearsals under his professional direction. I know the play was considered quite an accomplishment, and in mere length of time and endurance it became famous; for, though it started promptly at eight o'clock, King John did not succeed in emitting his last groan and giving up the ghost until the stroke of twelve.

I still have a vivid picture of Mr. Drum rushing at me out of a dark corner off the stage as I came off after having made an egregious blunder and threatening me with murder and sudden death if I ever did such a thing again. After that I thought my goose was cooked as far as he was concerned, but early in my senior year he actually talked over with me whether I preferred to play Macbeth or Henry V. I do not know whether he decided upon Henry V. because I liked it better or because he wanted to put on a rather elaborate scenic production of an historical play. At any rate, Henry V. it was and he went at this production with the same vigor and abandon with which he had produced King John.

I realize now how much of himself he gave to me, how earnestly he worked to bring me up to something like the standard he had set for himself. I remember that before the second production of Henry V. he came to me one day and said: "N. . . . . , you were a great distraction to me during my morning meditation today. I do not like the way you did that soliloquy before the battle of Agincourt." And he took me into the Academic Hall and we went over that one soliloquy for an hour and a half.

Nothing seemed to be too much of a strain or too much of a drain on his powers to obtain the effect he was aiming at. During the year he produced King John for the college, he put on a splendid production of "The Comedy of Errors" for the Academic Department, and he borrowed for one of his Dromios a member of our class—the present distinguished Reverend Charles Connor, S. J.,—and as though two performances of Henry V. were not enough for one year's work, during our senior year he produced for the Academic Department a really remarkable performance of "The Bells," having borrowed one of the members of our class for Matthias, —Tom Brennan—who died ten years after his graduation, then a New York Senator. He certainly had a way of winning support from the fellows at college, particularly the older ones. They were all willing to help with any academic production as stage manager, etc., and I remember that in "The Comedy of Errors" I helped him with the music.

Another side of his character was revealed in a little incident that occurred while he was preparing King John. I had a bad cold and was kept in the house for several days and although I lived in New Rochelle, he came all the way out there on Thursday afternoon to visit me and cheer me up.

Undoubtedly, these contacts I had with Father Drum in his efforts to make college boys produce dramatic masterpieces with real appreciation had a strong and lasting effect upon me and all the others who came under his influence. The educational value of such work, as he directed it, is incalculable,—it cannot be reduced to a formula or indicated by an index number, but it lives and grows and continues to be a delight and an illumination to the mind. I feel sure that boys who came under his teaching in the classics must have had the same reaction that I had in my dealings with him as a director of dramatic productions.

After I left college he went to make his Theology and I saw him seldom. I remember that the last time was a chance meeting in the Grand Central Station and we had only a few hurried moments to shake hands and renew old times. Now that he is gone, I can go back, as I have done in this letter, for more than twenty years and still keep an undimmed impression of an earnest, charming, and active personality,—a man, who, because he was himself an ardent flame, was able to hand on fire to others. May he rest in peace!

In September of the year 1900, Mr. Drum was sent to Gonzaga College, Washington, D. C., to teach the Freshman class. One of his boys, now a Jesuit, remembers well the thoroughness and painstaking exactness with which he did all his work: "Mr. Drum was rigid in marking, and demanded much from the boys: and his course in Greek was one of the finest courses I have ever studied anywhere in any branch before or since."

It was during this year that Mr. Drum first came into close contact with Father William O'Brien Pardow, who was the Spiritual Father at Gonzaga College, and preacher in the Church of St. Aloysius. Mr. Drum would later often speak of his admiration for Father Pardow, and how he made it a duty for himself to be on hand to hear him preach on every possible occasion, so as to familiarize himself with the style of the distinguished orator and imitate his methods.

At Gonzaga, Mr. Drum was given a task which was again to test his ability to handle boys and his capacity for organization. He took charge of the altar-boys, and within two months produced a veritable transformation in the affairs of the Sanctuary Society of St. Aloysius' Church which is attached to the college. Very soon the boys of St. Aloysius' were known everywhere for their reverence when on the altar and for the beauty and orderliness of their ceremonial. With his dignified, military bearing and his own deep reverence for the Blessed Sacrament, it was not a difficult task to instil into their young minds the same spirit of respect for everything that pertained to the altar. His influence was destined to produce permanent results, for the organization has since that time been most fruitful in vocations, and not a few of these are certainly traceable to the quiet, unobtrusive piety of their director.

One may still see in the corridors of the College, near the

main entrance, a group of large framed photographs, in one of which appears Mr. Drum and his Sanctuary Society, and of the thirty or forty boys in the group, some ten or more are known to have since been raised to priesthood.

Under Mr. Drum's supervision, the ceremonies of the altar were carried out with great splendor and with a certain air of military precision, on all the great feasts and at the High Mass on Sunday. In later years he would remark with pardonable pride that he had nowhere seen the effort made, on such a large scale and Sunday after Sunday, to enhance the glory of Catholic ritual, save only in the Jesuit Church in Beirut, Syria.

> I was very much pleased to find how well my boys carried out the instructions I gave them [he wrote later on in the diary of his travels, when he had occasion to note the conduct of altar-boys in many different churches.] The result was that they looked upon their position as a privilege, and not as a task. You see I'm a dreadful crank about altar-boys.

Father Drum's reading was voluminous. His notes contain evidence in abundance of the extent and comprehensiveness of his literary attainments. As was customary in his day and generation, he filled whole books with extracts from the greater poets and prose writers. "Must read and study well" is one of his remarks in a scrap-book, "all the better authors, Milton, Shakespeare, etc.; must not waste time on the others." On one occasion he said to the present writer: "There is so much said, and written, about Platonic "ideas," but I have read all of Plato in the original to find them; and have never yet come across any Platonic ideas."

Anyone who has the patience to gather together such a store of literary selections as are found in the notes of Mr. Drum could scarcely fail to imbibe something of the richness of thought and language and emotion that may be gained from contact with the world's best minds. There is evidence to show that the following letter, written to a Religious Superior who had asked his advice on a list of books for a college library, contains the authors which he himself had read thoroughly, and that his advice is given from first hand and actual acquaintance.

11 November, 1918.

Dear Sister,-

. . . .

Now for profane literature. I am far behind the times, and gleeful so to be. Aside from the novels of Monsignor Benson and John Ayscough, none occur to me that are of much worth; and even they fall short of the old ideals. I make no mention of the classics in fiction, except that among them I include Stevenson's David Balfour, and its sequel,—in fact, most of his novels, although latterly he went in overmuch for the wild and weird. Here are some good ones.

Ruskin, Newman, Matthew Arnold—all, except the Parochial and Plain Sermons of Newman, and Sheepfolds of Ruskin, and one or two of Arnold's on religious subjects; Stevenson—Memories and Portraits, Virginibus Puerisque; Lamb, Landor, Hazlitt,—the latter two are at times all wrong, yet could be given to college girls who are forewarned; DeQuincy, bad in spots; Jebb, Classical Greek Poetry; Butcher, Greek Genius; Homer, Chapman's, Butcher and Lang's, Lang and Leaf and Meyer; Lang, Letters to Dead Authors, and various books on religion,—he believes in the monotheistic origin of worship; Isaac Disraeli, Curiosities of Literature. How will that do? You see, the literature of the English language is vast; and the world's literature is vaster. It would be well to have some good translations of Dante (Longfellow, Cary); of Horace (metrical: Lytton, Prout's Reliques); of Virgil (Dryden); of Tasso, Cervantes, Camoens, Lopez de Vega. The last three I read only in the original, and am not sure that there are metrical versions of them in English. For the rest, when we next meet, I shall be able to help better than in writing. As to being good, read Romans 5, 1-5. In the Greek it runs: "Having been justified by faith, let us retain peace in our relation to God through Christ Jesus, Our Lord. Through Him, by faith, we have obtained an entrance into that state of grace in which we now stand firm. So let us exult in the hope of the glory of God. Not only that, but let us exult in our trials. For we know that trials beget grit; grit begets reliability; reliability begets hope; and hope never disappoints. For the love of God is poured out in our hearts by the Holy Spirit that is given us." Is not that great? Merely to understand St. Paul is reason enough to learn Hellenistic. I studied Italian only for the sake of Dante; and now find him of very little joy to me. St. Paul fills the need not only of spiritual, but also of literary stimulus. He had not the chaste and chiseled ivories of speech to wield in Demosthenic style; but his was one of the most masterful personalities ever expressed in masterful straightforwardness and sincerity.

If indeed further evidence were necessary to show how unremitting and absorbing were his efforts at self-improvement, and to accomplish this purpose, how thoroughly he carried out the plan of reading to which he set himself, a glance at his method, and manner of cultivating a fine literary style will be amply convincing on this score.

As we have seen, most of his earlier notes were torn up at some time or other before his trip abroad. But there is, among other records pertaining to this subject that somehow escaped the general destruction, a small notebook in the possession of one of Father Drum's friends, in Washington, which contains a fairly complete list of all the books that he managed to peruse in the course of a year or more, with the date of each book read, and after each name is added a critique of the author and the book itself. It affords a revelation of the process, as it were, by which Father Drum acquired his finished literary taste and his wide acquaintance with all branches of English literature. Again, it may be noted that all these books were perused during the summer months for the most part, and this is but another indication of that singular industry which would not allow him even to relax his active efforts. The list and some of the dates are here given, but space will not permit of more than an occasional quotation from the actual critique that follows the name of each book.

During Villa, Keyser Island, 1899.

Jane Eyre, C. Bronte; Anne of Geierstein, Scott; The Secret of Fougerouse, Miss Guiney; Venetian Life, W. D. Howells; Pascarel, Anon.

Aug. 5, 1899. Ernest Maltravers—Lytton. No matter how Bulwer insists on his good intention he certainly has written a harmful book, one in which vice is palliated. Maltravers is a sinful, weak-willed man, not at all extraordinary or attractive to me. Alice should have more strength of character.

Aug. 6, 1899. Alice—Lytton. In this sequel to Maltravers, my former convictions have been strengthened.

There is no one to love in these works; nothing to relieve the manifold vein of immoral love. Alice should have, in her life of sorrow, learned enough of human life and things above, to see how unworthy of her love Maltravers was.

Aug. 7, 1899. Rab and His Friends—Browne. What a relief from Lytton's sink of ill-placed love! How this simple, old, classic tale went to my heart! Strange that I never had read it before! The whole-souled love of husband and wife, the wonderful sympathy of Rab, the simplicity with which all is told,—these qualities have deeply impressed me. They are the qualities that draw us to F. Hopkinson Smith's "Tom Grogan," "Captain Joe," "Jonothan," etc.

Aug. 8, 1899. Marjorie Fleming—Browne. Such an interest did Rab and His Friends excite in me that I took up Marjorie Fleming with the fullest expectation of simplicity and love. Dear old Mattie Scott with his fond love for little Maidie, and his "onding a snaw,—aye, onding, that's the word." All this is delightful. What a knowing little one this Maidie was, with her talk about young bucks, etc. She had a very light vein in her *Grand Serieux*.

Aug. 9, 1899. The Culprit Fay—Drake. There is a Swinburne-like jingle in this poem, with all its coloring . . . old, old words and accurate musical flow.

Aug. 15, 1899. Sorrows of Werther—Goethe. [A book which is indignantly condemned.]

Aug. 16, 1899. Theodosius and Constantia—John Langhorne.

Aug. 18, 1899. The Castle of Otranto—Horace Walpole.

Aug. 18, 1899. Rienzi—Miss Mitford.

Aug. 19, 1899. The Flood of Years, etc.—Bryant.

Aug. 22, 1899. The Girlhood of Shakespeare's Heroines—M. C. Clarke.

Aug. 23, 1899. Avatar—Theophile Gautier. [This meets with a severe condemnation.]

Aug. 23, 1899. The Venus of Lille—Prosper Merimée.

Aug. 24, 1899. A Noble Sacrifice—Paul Féval. So did Féval attract me when as a Novice I pored over the charming pages of his "Les Jesuites" that I was very much afraid a novel by him would make to remove some of my first love. But this story is most noble and inspiring! M. le Baron de Kerlman is a most noble brother and lover. His love for Reine de Montmeril and the strength that put aside so great love from a sense of duty to Roger Saint-Magnon, his wild and hot-tempered brother, were such as would be looked for in a thoroughly manly chap, an out-and-out honest man such as M. le Baron Bertrand truly was. The greatest sympathy is aroused in the reader, and he feels pained that the noble sacrifice of love for Reine and for life itself is to be fully accepted; but in the end the dénouement rewards the noble, faithful love of Bertrand, and make a first-rate valiant fellow of the traitcr and fickle Roger.

Sept. 23, 1899. The Gadsbys—Kipling. Really, after having last year read Plain Tales from the Hills, The Light that Failed, Soldiers Three, Phantom Rickshaw, Barrack Room Ballads, Recessional, White Man's Burden, Seven Seas, Captains Courageous, and the Lord knows only how many of Kipling's peculiar outputs, I never expected to take up another book written by the same much-read author. But, the Gadsbys so attracted me that I could not give it up. And yet there is not any great art in it.

Sept. 23, 1899. The Late Mrs. Null—Stockton. [There is a blank after the critique of this book, and then the following remark.] I have failed in my plan of putting down these jottings. Shall try to be faithful.

July, 1900. "Fielding's Works." As I did not go to Villa, but worked in our library at St. Francis Xavier, I had much free time for reading.

Aug. 1900. Sterne's Works.

Sept. 2, 1900. Life of Nelson—Southey.

Sept. 3, 1900. Via Crucis—M. Crawford.

Sept. 24, 1900. Smollett's Works. [At the end of a long review of these and the preceding volumes he has this to say, of his short excursus in the field of post-Reformation English literature—] I must say that I have grown tired of poring over these time-honored, almost forgotten tales of two centuries and a half ago. All of them were built up in the same stereotyped form . . . Now for some more substantial reading, wherewith to feed the mind and curb the fancy.

After this I read Isaac Disraeli's Curiosities of Literature, and many of the old French dramas—works of Hardi, Rotrow, and their contemporaries, in the original. This reading was becoming very interesting and elevating, when the work of ceremonies and Altar Boys in Gonzaga began to switch me way off the line of serious study.

July, 1901, during Villa, Keyser Island.

Hugo—Les Miserables—With permission, I read and

greatly admired this wonderful analysis of human nature. What a pity that the innuendo against things Catholic, and the spirit of the Revolution cause the book to be a thing of danger. Great art does not strive for ephemeral notoriety by any sacrifice of truth.

Thackeray: Shabby Genteel Family, Adventures of Philip.

Dickens: Sketches by Boz, Short Stories.

Austen: Pride and Prejudice.

Aug. 10, 1901, at Woodstock College. Aylwin—Theodore Watts—Dunton.

Aug. 20, 1901. Wm. Barry—The Two Standards.

Aug. 21, 1901. Wm. Barry—The New Antigone. [The last remark in the critique of these two books is] I must let such trash go now, and dig into deeper soil for choice metal.

Aug. 28, 1901. Ruskin, Sesame and Lilies. I had often read this series of three lectures in part; yet now enjoyed a thorough perusal of them. In Sesame, the author speaks of books and reading,—"bread made of that old enchanted Arabian grain, the Sesame, which opens doors—doors, not of robbers' but of kings' treasures." Ruskin gives very ennobling lessons on what to read, and what not to read, on books of the hour, and books of all times. In Lilies he tells young ladies how they may be queenly, how they may joy it in "Queens' Gardens." The heroines of Shakespeare, Scott, and other writers are studied and analyzed very carefully. In the third lecture on the "Mystery of Life and its Arts," a number of favorite hobbies on religion and morality are ridden most intrepidly, and here and there some principles of art are discussed. I have been more than ever impressed with the wonderful simplicity and grandeur withal of Ruskin's style. What a noble mind in a wilderness of rank ignobility.

Oct. 7, 1901. Ruskin, Crown of Wild Olives. "It is wonderful how in his talks with the ordinary people, this master of style expresses himself almost wholly in monosyllables."

Oct. 13, 1901. Ruskin—Ethics of the Dust.

Oct. 24, 1901. Ruskin—The Eagles' Nest. These are ten lectures on the relation of natural science to art, given before the undergraduates of Oxford in 1872. Ruskin disapproved of the study of anatomy by artists, and considered that the polluted work of Montegna and Dürer, the body drawing of Botticelli and Holbein were failures artistically because of anatomy. In lecture II. we find the reason of the title of the book: men should not grope for science in the pit with the mole, but should build their thoughts high as the eagle does her nest. These lectures are teeming with the high-mindedness so thoroughly admirable in dear old Ruskin and contain little of the self-sufficiency that made him so many foes. The story of Halcyone is worked up, from divers authorities, most beautifully in the ninth lecture; chivalry and heraldry are treated in the tenth, and lead to a grand appeal that the young men of Oxford be not scoffers at whatsoever things are good and pure in the ages that so throve on love of Christ and His Mother.

Feb. 25, 1902. Ruskin—Frondes Agrestes.

Aug. 1, 1902. Scott—Talisman.

Aug. 6, 1902. Sienkiewicz—Without Dogma.

Aug. 14, 1902. Scott—Heart of Midlothian. These old-timers do give me pleasure. Of course, the Scotch are hero and heroine. Who can object to Sir Walter's leaning towards his aine? The breaking into Talbot prison, the Heart of Midlothian, by the Porteous mob, is well told. The plot is simple, of the old sort, but very interesting. Jeamie Deans is the noblest little heroine Scott has portrayed. What love of truth! What fidelity to her first love! What toil and never-ending efforts for her wayward sister's sake! Many a tear gushed from my heart at the happy turn of the denouement. The simple-hearted, whole-souled eloquence of Jeamie's words to Queen Caroline move one to deepest sympathy. How kind the girl was to wayward Effie, and to the old Cameronian cow-herd, Douce Davie Deans, father of the twa [*twa*=Scotch] lassies! The Duke of Argyle is painted as a true Scott, a lover of his braes and his braw lads and his gentle lassies, of his cheeses and snuff and usquebach, of all that concerned the land that he represented! One is much attracted to the Duke; but George Staunton, who was first a minister's lustful son, then a highwayman, then a nobleman and hypocrite, excites no sympathy. Scott's ignorance of Catholicity makes this living lie to join the Church which "pretends" by maceration of the body to expiate the sins of the soul. Effie also becomes a Catholic!!

Aug. 28, 1902. Crawford—In the Palace of the King.

Sept. 5, 1902. Sienkiewicz—Children of the Soil.

April 30, 1903. Tyrrell—A Handful of Myrrh.

May 20, 1903. Faber—Spiritual Conferences. These Conferences have pleased me very much. Their aim is very high and must have been intended for people who are striving earnestly for perfection. There is, in Faber's chatty exposition, a freshness and freedom from the commonplace, a crispness and nicety of diction, that charm the intellect with what are the time-worn truths of ascetics. Golden sayings are many and new in Kind Thoughts, Kind Words, Kind Actions. Really, Faber spurred me on to be kinder. It is wonderful how much theology he crowds into his pages and that, too, without overcrowding things, and getting away from the style that pleases people in him. In his conference on the special vocation that all men have, he surprised me with his rich amplification in describing the wonderful power and number of actual graces we receive. The theology of the Society was set forth without any mention of the controversy. This knack of knowing just how much of theology to serve out, and the when and the where of the serving, is a very important consideration for the preacher of the word of God. Faber's use of homely idioms, telling illustrations, simple language, rapid thought, uninvolved structure, is noteworthy.

July 22, 1903. Kennedy—Rob of the Bowl. The author seems to have made an accurate study of the country around about St. Inigo's. Much of the plot is transacted in the Rose Croft, just opposite our Villa. St. Inigo's Creek, our Point, and the stretch of the water off from the Villa are ever recurring in the story. The historical accuracy of the tale is rather doubtful. Indeed, apart from the interest that arises out of the locality of the plot, Rob of the Bowl is not worth much.

July 22, 1903. Bulwer—The Caxtons. Here the strongly portrayed characters are sure to teach one a salutary lesson in the natural virtues needed for life. Noble, virtue is ever triumphant over vice and sin.

Aug. 11, 1903. Bulwer—What Will He Do with It?

Aug. 13, 1903 Eliot—Theophrastus Such.

Aug. 15, 1903. Ravigan's Last Retreat.

Aug. 16, 1903. Eliot—Poems.

Aug. 25, 1903. Vicomte Henri de Bornier—Romance of a Playwright.

Aug. 31, 1903. Wilhelm Walloth—Empress Octavia.

Sept. 11, 1903. Thackeray—Barry Lyndon. What a treat it is to read Thackeray's keenest satire and get down into his kindliest heart! When he quotes the dear Old Heathen there is so much of the rollicking spirit of Prout and so little of the pedantry of Bulwer that the quotation refreshes and adds a gleeful measure to the jolly piece of wit. England's duty to down Ireland, and the certain unfitness of the Irishman to do anything well, are occasions of many witty and satirical cuts in Redmon Barry's Autobiography.

Aug. 16, 1904. Eliot—Daniel Deronda. Only one novel for this vacation! Good for nothing fiction palls on me; the great novels carry me along too much and too far for present occupations. This is the only one of Eliot's that I had not read. It held me and preyed upon me.

CHAPTER VII

WOODSTOCK AND THEOLOGY

In September, 1901, we find Mr. Drum again at Woodstock, this time for the study of theology. His course, however, was not to be as placid as in the earlier days of philosophy. He at first resolved upon a strenuous program of private study and reading, and, besides attending with his usual thoroughness to the work of the ordinary classes, mapped out for himself a course of apologetic writing, and undertook also to prepare several papers for publication. It is true his talents were equal to these tasks; but from the very outset of the first year, he was troubled with ill-health. He suffered severely from poor digestion and was annoyed by incessant headaches, and the indisposition was at times so serious that he found it impossible to remain at his desk and study for more than half an hour at a time.

But he did not complain, and few realized the difficulties under which he was laboring. For one thing, his private meditation notes of this time reveal that it was inherent loyalty to the cross that alone sustained his vigorous spirituality, and the glory of the priesthood shining in the near future was an ever present encouragement to him to plod sturdily on through the first hard winter.

Towards the end of March, however, it was necessary to send him to Washington for a two weeks' rest and for medical treatment. This vacation restored him considerably, and he had little difficulty in finishing out the year. In fact the college diaries indicate that he was appointed to take the defense at the Spring Disputations at Woodstock, in the treatise *De Gratia Christi*. But in the following years physical hardships continued to dog his footsteps, and he remained rather weak in health and frail in appearance even up to the end of his course of studies. But in spite of these difficulties, he managed to do exceptionally well in all branches of study. He once told a close friend about his examinations, that invariably he would pray beforehand to fail in them, if that would be for the great glory of God. One might say that this was rather a diplomatic approach to the source of inspiration, though a courageous one, and in point of fact he never did fail. The prayer is far different from the pleadings that usually ascend to good St. Joseph of Cupertino, or other patrons of examinations at the time of the last academic ordeal.

The handicap of ill-health did not prevent him from accomplishing a deal of labor that might tax a more robust constitution. It speaks well for his strength of will and energy, that he actually succeeded in preparing and publishing several papers of his own, as well as in arranging a plan for a series of scholarly papers by the "Theologians' Academy," a volunteer organization among the students for private study and research. He was elected president of this academy for a year, and afterwards succeeded in having twelve of the papers published.

In his teaching days Mr. Drum had been most successful in the study of Greek. His notes are bewildering in their abundance and show the evident painstaking care he must have taken. He gave perhaps more time to this language and to Greek literature than to any other part of the course, and it appears that he really produced good results with his classes. During theology, he took some of his spare time to gather together the fruits of his careful study, and wrote two lengthy essays, one on "The Origins of the Homeric Religion," and the other on "Traces of Revelation in Homer." These essays he afterwards combined together, under the latter title, and published the result in the *American Catholic Quarterly Review*, April, 1904. The article is twenty-one pages long, and is a careful study of the religious ideas of the Homeric and Mykenaean age. The same magazine also printed a later paper, on Eddyism, under the title of "The Meaning of Christian Science," in the issue for January, 1905. At about this time he was invited to collaborate with Dr. Alexander E. Sanford, M.D., in the revision and re-editing of his book on "Pastoral Medicine," to which he contributed the chapter on "The Moment of Death." The book was republished by Joseph Wagner, N. Y., in 1905, and made his name a familiar one among the American clergy long before the period of his more active writing on theological subjects.

Besides this literary work, he was also asked to conduct an advanced class in Greek for the philosophers. He had also, like everyone else, to write one or two sermons, to be delivered at the noonday meal, but besides the required number, there are in his notes several long sermons all dated at about this time, so that he must have shown an unusual amount of industry and fidelity to writing. Of course, too, it was to be expected that such a man would be faithful to that perennial duty which was the great burden of his day and generation and long afterwards: the Latin poem which had to be written twice yearly by every Scholastic. One of these poetic and classical effusions of Mr. Drum, written in honor of St. Aloysius, contains thirteen stanzas in the Sapphic meter, all fairly rhythmical and accurate.

His voice, too, had been carefully attended to, and had developed into a fine baritone, which appeared to advantage on several occasions, notably at Woodstock's celebration of the Jubilee of the Immaculate Conception on November 6, 1904.

About this time, Father Drum prepared a paper on the Jesuit observatory at Belen, in Havana, Cuba, which won public recognition and deserves at least a passing notice at this point. His attention was called to the subject quite accidentally, and his quick reaction to the occasion was entirely characteristic. Happening to see a report of the U. S. Weather Bureau, in which an account was given of the opening of an American branch office in Havana, his attention was aroused by a singular statement at the beginning of this account, which informed the public that

> at first it was difficult to interest the Cubans in the warning service, since they are by nature very conservative and slow to adopt any change in their accustomed methods and mode of living. The issue of warnings of hurricanes was a radical change, the inhabitants being accustomed to hear of these phenomena only upon their near approach.

Father Drum's interest in Cuba had already given him enough knowledge of the country to suspect the falsity of this statement, and he began to investigate the full facts. He procured several publications of the Jesuit meteorological observatory at Belen. The results of his study soon appeared in an article in the *Messenger*, for June, 1905, under the title "Pioneer Forecasters of Hurricanes." He presented the facts to show that for fifty years the Cubans had been receiving regular service from this observatory in the shape of weather reports and storm warnings of a most reliable character, and recounted the signal achievements of such internationally known scholars as Fathers Viñes and Gangoiti and other Jesuit meteorologists. Even at that very moment the observatory was publishing studies and reports that fully justified its international reputation. A few years previously, in September 1900, they had published forecasts which, had they been heeded, and had it not been for the contradictory forecasts of the U. S. Weather Bureau observers, might have done much to lessen the disastrous effects of the terrible Galveston storm. In the face of such facts it was impossible to justify the ignorant criticism that had been launched in the report. The tone of Father Drum's article on "The Pioneer Forecasters of Hurricanes" was so vigorous and the array of facts so convincing that the pamphlet was asked for by the Chief of the Weather Bureau, Prof. Willis L. Moore, who had it reprinted and published by the Weather Bureau, and then gave orders for the closing of the U. S. Weather Bureau station at Havana. He requested Father Gangoiti to send daily cable reports to Washington, at the Government's expense, during the hurricane season. Since that time there has been a continuous service from Belen, which sometimes sends to Washington many reports in a single day.

Professor Moore expressed his personal thanks to Father Drum for calling his attention to the real state of affairs in Cuba, and needless to say, the latter felt amply satisfied with the result of his sturdy vindication of the Jesuit observatory at Belen. He was invited, in the following year, to accept an honorary membership in the Hispano-American Society of Washington. The newspapers of Havana gave the pamphlet an enthusiastic review, one of them, *El Comercio*, gloating over the fact that "the compatriots of Roosevelt were proved at last to be inferior in science, even if they were superior in war."

The great day of Ordination finally drew near, at the end of the third year of theology; and on June 28, 1904, together with eleven other Jesuit Scholastics, Mr. Drum was raised to the sacred priesthood by his Eminence James Cardinal Gibbons. The Ordinations took place in the little chapel at Woodstock. His mother and three of his brothers and a number of close friends were present for the ceremonies, and for the solemn splendor of the following happy morning, on the feast of Sts. Peter and Paul, when for the first time, the newly ordained Priest exercised his privilege of offering up the Pure Oblation, and with his own hands gave Holy Communion to his mother and happy family.

Father Drum had always been a man of very high ideals and always thoroughly in earnest in his vocation. The priesthood and its closer contact with the altar that he loved fanned his zeal to renewed intensity. There is a little quatrain embodying a thought of Fr. Olivant's, which he copied into a notebook at some time during his last year of theology, and which he even translated into a Greek distich in honor of one of the newly ordained priests of the year following his own Ordination. It seems true to say that the thoughts and aspirations of his whole life at this period were crystallized into these brief lines. The original form of the quatrain seems to have been in the second person, but it reads as well in the first, in the form which he used on many occasions in retreats:

His priest am I at break of day
Till that my sacred task be done;
As for the rest, 'tis He holds sway:
His victim I, till sets the sun.

That for him this was no idle sentiment is clear from many incidents that occurred in this and the following years to test his spirit of obedience. There was one particularly trying experience with the censors, who rejected an article on which he had spent much time and labor, and which he was not allowed to publish. "This is the time," he wrote simply, to a friend who had helped him with the article, "this is the time for renunciation of will and judgment."

The study of theology, in the one year that followed Ordination, took on a new meaning. His one absorbing delight, then and for a long time to come, became the reading and study of St. Paul. He often said later that he then began to study St. Paul with a passion, and took up the Hellenistic dialect in order the better to appreciate his true meaning. That he made frequent and vigorous use of this predilection for the great Apostle was manifest even in the year of Tertianship, and in his preaching in later years. Perhaps it was this study more than anything else that helped to deepen his already vivid faith and clarify his spiritual aims. It is certain that from this time on he became more than ever militant in the cause of the Faith and of Catholic truth.

However, there are remaining from this period only a few notes that bear on his meditations and spiritual exercises. It must have been towards the end of his last year of theology that he tore up many of his earlier notes; yet some excerpts remain that enable us to see something of his advance in the science of the saints.

One might readily conclude, however, from the narrative so far, that Father Drum's personal spiritual affairs were not suffered to be neglected. We have seen that he made a continuous record of his meditations, although the record is broken by long gaps, especially during the teaching period, very likely because the ordinary labors of a Scholastic in the Regency were sufficiently strenuous to take up all his time and attention, and allowed of little leisure for taking notes for his own future use. But at least for the summer months the journal seems to have been kept with consistent fidelity, and something is put down each day to preserve the thoughts and the lights he received in prayer. A few selections are here given, chosen more or less at random, to indicate the intensely practical nature of his meditating, and how serious were his efforts to maintain the high spiritual ideals of his earlier Religious life.

> Aug. 15, 1900. The Assumption! Our Lady chose to remain away from the glory of her son during the ten years that intervened between His Ascension and her assumption into heaven, because of her very great zeal. How did she show this zeal? In her attitude towards God? towards men? How did she the Christ work? How do I imitate her?
>
> 2. The glory of the Assumption was a reward for labor done. It was a carrying out of the promised victory of the Kingdom—a victory in proportion to the toil of battle.
>
> 3. Contemplate Mary's glorified body—its clarity, agility, subtility, immortality. *Haec est vita aeterna! Quam sordescit mihi terra!*

Aug. 21st. After the temptation, Luke iv, 1. And when the temptation was ended, the devil departed from Him for a time. What a true picture of the earnestness and indefatigability of Satan! Departed for a time! Such action as that of Jesus should have prostrated the devil, and rendered him void of hope; but it made him depart only for a time! What do I?

Aug. 24th. Call of Peter and Philip. 1. Peter follows the good lead of Andrew and is brought to Jesus! Did he ask proofs of Andrew? His simple obedience was mightily rewarded! Jesus immediately treats Peter in an extraordinary way. He changes the name Simon to a name that always gives a clue to Peter's future. Of course Christ saw the future; but as Peter came up to Our Lord, that plain, outspoken honesty of his must have pleased Jesus.

Aug. 25th. 2. Philip—Wonderful personal influence of Jesus! Follow me! How often have I in retreat after retreat, day after day, heard the inner voice calling me to follow Jesus.

Aug. 27th. Call of St. Matthew, Luke v. 1. Follow me! What simple orders! How well carried out! No one thus called fails. Levi leaves all things on the table, rises and goes with Jesus. What a magnetism in the person of Our Lord!

2. Levi prepares a feast. Jesus eats and drinks with publicans and sinners. No caste separation hems Him in and offsets His good work.

Sept. 1. Sermon to the Twelve (continued), Luke vi. Love your enemies, do good to them that hate you. How different from the principles of worldy wisdom! Did not Jesus carry out this principle and bless those that cursed Him, and pray for those that calumniated Him? Of what use is worldly knowledge, popularity, applause, if they withdraw one from the true principles of life?

Sept. 5. Raising of the son of the Widow of Naim—Luke vii, 1. Compassion of Jesus for the widow bereft of her only son. Weep not! See Him, examine His thoughts, hear the words. Compare with this scene many occasions wherein He comforted you, or would have done so had you shown faith in Him. He went to the widow. He touched the bier. He could not endure to see the woman so grieved.

2. The miracle: "Young man, I say to thee, Arise! And he that was dead sat up, and began to speak." Wonderful power, love, kindness of Jesus! How much He would do for me, did I allow of it.

3. The mother's joy. The fear of all! They now glorified God, for that they feared His awful might! How short a time did this mighty fame last! Why did not Jesus now declare Himself King that He was?

Sept. 12. Jesus in the House of Simon the Pharisee. Contemplate the women, probably Mary Magdalene, entering and pouring the ointment on the feet of Jesus. Think her thoughts. What did the Pharisees think? This man, if He were a prophet, would surely know who and what kind of woman this is that toucheth Him." *Attende tibi et doctrinae*! Follow St. John Berchman's *attendam serio quid mihi agendum sit, non quid agatur ab aliis*. There is a great deal else for me to do than to wonder why God allows thus and so!

Sept. 15. Jesus' Mother and brethren come to see Him. Luke viii. Nowhere before was mention made of Mary's having called on Jesus during His public life. She met her Son at the marriage of Cana, but this meeting was accidental. Now, He sends answer to those who are without: "My mother and brother are they who have the word of God, and do it." What a consolation for me who am called on to hear and do God's word! How intimately I shall be classed with Him!

Sept. 20. Raising the daughter of Jairus, Luke viii. Jesus goes at the first request of the leader of the synagogue; He is ever ready to do good. Even while intent on aiding one family, Jesus is sought after by others. What great faith in the poor woman with the issue of blood! She only wished to touch the garment of her Lord! He knew the touch of faith! He needs not that we tell Him and remind Him of our call on His aid. No, it is the spirit of faith that is wanting in us. "Daughter, thy faith hath made thee whole; go in peace!"

Sept. 23. Address after the miracle of multiplication, Luke ix—"If any man will come after Me, let him deny himself and take up his cross, and follow Me." What a hard doctrine to preach after so kind an act! Jesus never fails of laying before the people His doctrine in all its sublimity. They must not expect the following of Christ to be all wonders and no hardships; they must follow the cross, yea, bear the cross! "For whosoever will save his life shall lose it; for he that shall lose his life, for my sake, shall save it." What indeed does it profit a man to live and be happy here on earth, yes, to gain the whole world and all the natural joy thereof, if he lose his soul and lose the everlasting joy of heaven.

Aug. 24, 1901. St. Bartholomew, Luke vi. The call

of the twelve. 1. Christ went up into the mountain to pray, that He might save His enemies by the apostles. He wished to save by light, not by power; by the medium of His grace and His apostles, not by the will to bring us to heaven perforce; by the transmission of His grace through the humble fishermen whom He chose to represent Him.

2. He nighted it in prayer with God.

Aug. 25, 1901. Most pure Heart of Mary. 1. God made it full of grace. There was not even original taint or tendency to sin in it. The Mother of God must have a heart filled with love to God. How sad the state of my heart! Yet, "whosoever doeth the will of my Father, he is My mother!" He indeed will be filled with great graces of God. What are those graces?

2. It was most devoted to men. *Ecce filius tuus!* From the statement of those words, man has been the special concern of Our Lady's heart. Yet how could we have been! What return, to the *Ecce mater tua!* have I made? Our Lord's words to Mary and to us remain in force. Let me then turn to that sweet Heart, and pray its special pity and concern for me!

Aug. 27, 1901. The Growth (Mark iv, 1-20). The supernatural growth of virtue is imperceptible—like the natural growth of seed. Man sleeps by night, rises by day; leads his busy here-today-gone-tomorrow kind of life, and the seed he has sowed gives forth the blade, unless it be harmed. It is necessary only that the soil be fitting. When is my heart fitting soil for virtue? When detached from things of the world, the flesh and the devil? In such soil, virtue springs up and is strengthened while man knoweth it not.

Sept. 3, 1901. Jesus condemned to death. Who was Pilate to condemn Jesus? What did Jesus when condemned? *Tacebat!* The silence of Jesus should be an object of faithful meditation. For an hour of small talk, of worldly talk, of nonsense and worse than nonsense does great harm to Religious life. Who was condemned? Why? With what results? To the Jews? To us? To what was he condemned? *Propter quem?*

Sept. 8, 1901. Nativity of Blessed Virgin Mary. Mary is born for the sake of first, God; second, us; third, her own salvation.

1. She was born to love and serve God as no ordinary creature could. To give birth to God in human flesh! To raise God-Man! To sorrow with Him, work for His work!

2. She was born for our example, salvation, sancti-

fication. By her mediation was salvation brought into this world; by her womanhood lifted up! By her intercession have countless graces come to me.

3. What grace was hers, if at the moment of the angelic salutation, she was full of grace! How that grace increased, multiplied! What in my case? I am today a Jesuit eleven years!

Sept. 9, 1901. St. Peter Claver. 1. Preparation, graces received. Youth and opportunities sacrificed. Fidelity to grace! Hard study! Fervent prayer!

2. Service: its kind and its manner. To whom? How given? How rewarded? Example of servant that tended the sick Claver, abused him, ate his food. Claver's care for the ulcer-stricken! His wonderful cloak: ever used to cover and comfort the negroes reeking with sores, yet ever clean and sweet! God's grace.

Dec. 18, 1902. Matt. xii, 39-43. Blessed Sacrament—our neglect. Men of Ninive will rise up at the day of judgment, together with this generation and shall condemn it. For they did penance at the voice of Jonah. Lo, you now, there is in the midst of us a greater one than Jonah, and we hear him not. The Queen of the South shall likewise rise up with us on the day of final judgment, and shall force us to strike our breasts with sorrow, to hang our heads with shame at our so great neglect. She journeyed from the farthest lands to hear Solomon. Lo, who is it that we may hear?

Dec. 24, 1902. Vigil of Christmas. *Hodie scietis quia veniet Dominus, et salvabit nos, et mane videbitis gloriam eius.* Years and years the holy ones of the Old Law waited. They knew the Lord would come; they knew their need of Him, not so fully as we do, but fully enough to yearn for Him that should come to bring them salvation.

Jan. 23, 1903. Luke iv. 1. "He was led by the Spirit into the desert forty days to be tempted by the devil" (quoted in Greek). It would seem from the original that Jesus was for forty days led by the Spirit into the desert and was all along tempted by the devil, not that he was led thither, and then tempted forty days, but that the more the temptations the further he went into the desert throughout the period of trial. If there be any right to look at the Gospel story in this wise, it presents me with a fitting antidote to temptation: go further into the desert. That desert now means prayer, self-abasement, the *ama nesciri et pro nihilo reputari.*

CHAPTER VIII

EN ROUTE

While at Woodstock, certain of Mr. Drum's friends had agreed together to urge him as much as possible to the study of Scripture and of Oriental languages. His thorough mastery of Greek and his talent for language studies in general made it plain that he might in all probability be successful along these lines as well. He seemed to be destined for the Woodstock faculty in any event, and there were some who were perhaps a bit afraid of his dogmatic turn of mind and would have preferred to see him steered off from the teaching of philosophy and theology.

Be this as it may, Mr. Drum found it to his own inclination to take this suggestion, and began to devote himself seriously to Scripture work and to the mastery of Hebrew. His notes on the latter subject, taken at this time, are as thorough as usual, and form a brief grammar of the language. A fellow student remembers that he made it a point to write out a long Hebrew sentence each day. But when the time came to make selection of a man for the course of Holy Scripture and of Oriental languages, he himself was scarcely consulted about the matter, though everyone else was of opinion that he was the most likely man for the course. Fr. Timothy Brosnahan, then Prefect of Studies at Woodstock, recommended him to the Provincial. Another Father whose opinion was asked, replied that there was no doubt of Fr. Drum's ability: he knew his theology well, he had an evident gift for languages, his health was fairly good and his constitution strong, and he would moreover very likely fancy such a career. But he himself was not informed of the plans in his regard until the actual decision was made and all arrangements had been practically completed. And then his acceptance was prompt and unquestioning.

In the beginning of September, 1905, he began his tertianship, as the Third Year of Probation is called, at the Novitiate of St. Andrew-on-Hudson, Poughkeepsie, under the direction of Father William O'Brien Pardow. He was made Beadle of the Tertian Fathers and thus came into more frequent contact with Fr. Pardow whom he had known as his Provincial, and also as Spiritual Father at Gonzaga College, and for whom he had early developed a great admiration.

We may judge of the intense earnestness of his seeking after the perfection of his vocation from the following excerpts from his notes of the Long Retreat, which he made early in this year under the direction of Father Pardow. They reveal the consummation of his desire of the old Novitiate days, one of the graces that he had marked down for his daily earnest prayer, the grace of complete surrender, of giving to God the complete sacrifice of a devoted will, by means of the vow of perfection which he was then permitted to make, in imitation of Ven. Claude de la Colombière:

> Sept. 30, '05. There is very much to do during these days of prayer. The next great event of my life is death, the end of life.
>
> I put myself wholly in the hands of God. May the S.H. be merciful to me, may Mary show herself a Mother, may holy Father Ignatius, SS. Francis and Aloysius and John and Stanislaus watch over me. I hope to offer a higher gift to my God.
>
> Oct. 1, '05. I am once again resolved to be indifferent. No matter whether I be sent to the Philippines or not. For years I have been desirous of going on the foreign missions. During the last few months the old time desire has cooled somewhat, due to the realization that some good is apt to come from my preaching in English. I will hold my soul in readiness. This notion about preaching may be only a fancy . . . I will apply once again for the Philippines. May God direct my superiors.
>
> Oct. 4, '05. Four times today I have made the *quartum exercitium resumendo* . . . The triple colloquy will, by God's grace, be more of a factor in my life, than it has been. Some of the world's views have come into my life, and led to disorder.
>
> Oct. 5, '05. The realization of my ingratitude is greater this morning than it was yesterday, and has caused me tears of shame and sorrow. There seems to have been a very special providence even in my birth. According to mother, God wished me to be born that I might become His priest. He has hedged me round with every

help of family and school and countless graces. And what have I done? What fruit have I borne Him? Weeds and leaves, not the rich clusters He meant me to produce.

Oct. 6, '05. During the midnight and morning meditations I was rather heavy and dullish. My head is fagged out. I have probably not cared for it. The method of St. Ignatius is so fearfully in earnest, that one must need be very prudent not to suffer harm by it . . .

I feel that life in solitude, in prison, in destitution would be far easier and sweeter to me than the life that I am called on to lead. I am to be crucified to the world, while the world flatters and fondles me. I am very much in the world, yet not of the world. It frightens me to think of the many dangers of the life ahead of me . . . unless God grants me superabundance of grace.

Oct. 10, '05. Had St. Francis Borgia made much of wealth, and rank and power, God would not have made much of him. My soul should make much of my Lord and only of my Lord. The best tribute I can think of is to offer to my Lord, through Mary, the vow that my soul yearns to make. He will strengthen me who made Mary great and strong.

Oct. 12, '05. The offering of the Magi was never taken back. I hope to offer to the dear little Baby a life that, by God's grace, I will not take back. Our Lord is giving me much and sweet consolation. Today, during Mass of our Martyrs, especially at the great epistle of St. Paul, I was quite overcome with tears. May God help me to remember the sweetness He has granted me. The life of faith is made more clear to me by the Magi. I shan't like to meet these three *insignes* after all my cowardice and infidelity.

Oct. 14, '05. My meditations on the Standards had to be made very quietly. There is no need of enthusiasm just now. Excitement and feverish brain work would do me much harm . . . My poor head is tightened and sore; the rush of blood is irregular and there are spasmodic brain-pains. The old brain-fag threatens me. So I shall have to give up the midnight meditation and do less penance.

Oct. 16, '05. God's grace is working in my soul sweetly but surely to prepare me for the work to come. I am resolved to imitate Our Lord's tactics . . . The Father's will for me is that I observe our Holy Father's will as I find it in our rules. *Voluntas S. Ignatii est voluntas Dei, et voluntas Dei est sanctificatio mea.* [St. Ignatius' will is God's will, and God's will is my sanctification.] This thought gives me much joy. The prospect of a life of prayer and steady earnestness is my hope against the discouragement that my present weakness and future temptations tend to bring me. For God's grace will be with me if I am prayerful and exact in religious duty.

Oct. 23, '05. The sentence is given against my dear Lord. He bears the cross and takes the insults that are showered upon Him, and all the while His heart is full of love for me . . . He wishes His sorrowful passion to strengthen me. Father Instructor has granted me permissions to take the vow I have so long yearned to take. May it bring me nearer to my sorrowing Jesus. May my life be one of contempt like to that of the *vita vera.* May the way of the cross be my way!

Oct. 25, '05. During this break-day, I am joyfully thinking of the sacrifice of myself during the morrow's sacrifice of Jesus.

Oct. 26, '05. This morning I prayed much to Bl. Margaret Mary and to Ven. Claude de la Colombière. They went to the S.H. with me and I offered myself to the Sacred Heart of Jesus and begged Him to accept the sacrifice that I was about to make of myself and of Him. I then went to Confession. My Mass was very helpful to me. Just after the *panem coelestem accipiam* I made my vow *mentally,* and then received the food that is to give me strength to keep that vow. During Thanksgiving, I made my vow *verbally.* The formula—that of the simple Vows, except that, before the words *a tua ergo* I inserted the words, *Insuper voveo regularum nostrarum observantiam perpetuam.* The meaning—that I bind myself under pain of sin to observe for the rest of my life the Summary of the Constitutions, the Common Rules, the Rules of Modesty and the Rules for Priests. [Follows a detailed resume of all the rules in order, indicating exactly how they are to be understood and practiced.]

This vow does not frighten me. I mean to go on cheerfully in the observance of the rules that St. Ignatius meant to be observed. I have often been tripped by self-love and the devil, but have really tried to observe our rules. I wish to cut away from self-love and from other creatures and to belong to God alone. My efforts to live the life of the companion of Jesus will now be easier than before. This vow means to me an increase of God's grace,—a new main wire from the great power-house to my soul. The vow means more earnest efforts on my part. That will be easier to accomplish, to which I give more earnest efforts and God gives greater graces.

To prevent scruples, I lay down the following guide-posts:

1. Violation of this vow requires full attention and deliberation.
2. I do not bind myself to anything about which I doubt.
3. I do not bind myself in those cases in which charity or some duty may lead another to dispense himself without doing anything contrary to perfection.
4. I shall always obtain general permission to do certain things that community life now seems to demand that I should do.
5. If I need something out of the ordinary, I do not take its use to be a departure from common life.

Now that I have bound myself by closer ties to Jesus, my will is strengthened mightily, and my feelings are quite overcome. I thank Bl. Margaret Mary and Father de la Colombière for obtaining from the Sacred Heart for me these tears and sweet consolations. They are all helps. Above and beyond them is the increased good will which I pray the Sacred Heart to keep within my soul, even after sweet emotions have given place to the unfeeling life of rush and activity. May God keep me in my firm resolves and increase my vigilance and my gratitude.

I have been very happy today, and firmly resolved to bring more actively into my life the motive of love for Christ. He is risen, and is now in glory. That should be enough to make me happy and cheerful.

My vow will help to choke self-love in me. Self must be put down. Christ must go in self's place. The joy of the risen Christ and His glory should take the place of earthly joys.

Oct. 29, '05. On this last day of the Exercises, I feel very grateful to the dear Lord for the graces He has favored me with. He has kept me in such health as to enable me to attend every exercise and has given me much light and strength . . . Such a season of great grace will scarcely be mine again. Now I must do my share, and remember that it is *gratia Dei mecum.* God really wishes me to be a saint, i.e. to be true to my vow, to keep our rules perfectly. I hope to be more devoted to His Sacred Heart as a result.

It is customary for the Tertian Fathers to go out and give missions during the Lenten season, in company with members of the regular Mission Band. One of Father Drum's assignments was for a mission in the Church of St. Agnes, in Brooklyn, N. Y. A friend who afterward became a Jesuit heard him preach this Mission, and was impressed by his tremendous power as a preacher even then. He was a slightly built man, dark, and with flashing eyes, but he had a most powerful voice.

> Father Drum told me afterwards [so this friend relates] that he did a most unusual thing, for him, during that mission. At the end of the sermon on mortal sin, he finished his peroration with an awful curse—"May your house burn to the ground; may all your children perish in the flames; may you lose all that is dear to you in the world," and so on through a long list of fearful imprecations, ending with the single brief conclusion "rather than that you should ever be guilty of mortal sin!" He found that the people were looking up, startled and gasping, waiting for the end. He never again repeated that curse.

Towards the end of the year of Tertianship, Father Drum was asked by the Provincial, if there was any special work he had in mind for the future. He replied that there were three separate plans over which he had pondered, any one of which he would be glad to follow for the rest of his life, if his Superiors so wished. The first plan was to volunteer for missionary work in the Philippines. The Province was then thrilling with the story of the heroic missionaries who had gone out to the Philippine Islands. Again, he had followed with absorbing interest the varied experiences of his brother, then Captain Hugh A. Drum, who served through the three campaigns in the pacification of the Islands; while his own extensive reading on the subject of the Philippines, and his fluency in the Spanish language all convinced him that this was an exceptionally promising career to follow for God's glory and the good of souls. The second plan was to devote his life to preaching, especially in the line of apologetic and dogmatic sermons and in the giving of retreats and missions. This plan was even more congenial to his active nature, and to the peculiar gifts of oratory and the dramatic ability that he could command.

Finally, the last plan, about which, to tell the truth, he was not at all as enthusiastic as about either of the others, was to specialize in Oriental languages and in Holy Scripture. He knew this career would entail a life of drudgery and research work to which he did not feel himself in the least attracted.

However, it was not till the middle of summer that the decision was made and the assignment given to him.

That he did not even at the end of the Tertianship have any definite knowledge of his future course is clear from the absence of all reference to it in the following list of favors to be prayed for, which he wrote in his note book at the end of the retreat for this year:

> Graces for myself, to be included every day in the Holy Sacrifice, and to be begged of the Sacred Heart of Jesus, through the intercession of Our Lady, St. Joseph, my Guardian Angel, Sts. Ignatius and Xavier and all my heavenly patrons:
>
> To reach the high perfection that God calls me to. To say Mass with fervor and devotion. To say my breviary prayerfully and carefully. To understand the Exercises, and to use them as very effective tools for the salvation of souls. To win many souls to the true church and to the state of grace. To preach in such a way as to hold the attention of the people, to convince and to teach them of Catholic truth, and to move them to Catholic life. To keep my vows perfectly. To be able to keep forever my Tertian-vow. To be truly devoted to the Sacred Heart, to bring many to such devotion, to work ever for the Sacred Heart and never for self. To be a loving son of my Blessed Mother. To obtain all the favors set down in my older list. To die fortified with the last Sacraments of the Church. To have very good health, and to work thereby very earnestly for the salvation of souls. To have an ennobling influence on others. To be very obedient. To keep from unkind words about anyone. To find no fault in my Superiors. To see Christ in my Superiors. To look upon all others as my Superiors. To be true to God's graces. To shirk no humiliation. To desire earnestly the insults wherewith Christ was borne down.

The Tertianship closed with this retreat, which ended on June 21, 1906. He left for New York, after having previously made arrangements with his mother and sister to meet them at the College of St. Francis Xavier; but on his arrival there, received an order from the Provincial to leave immediately, to give the Thirty-Days' Retreat to the Christian Brothers at Poçantico Hills. He had to start without seeing his mother and sister, who reached St. Francis Xavier's later in the day and seemed at first extremely agitated at not finding him there. Two weeks later Father Drum received word that he was to go to Syria for a course of studies in Holy Scripture and Oriental languages. The retreat to the Christian Brothers had to be given up after the twentieth day, and another Father was assigned to complete the Exercises.

While waiting for arrangements to be made for his voyage, Father Drum was placed on duty at the Metropolitan Hospital on Blackwell's Island, now Welfare Island, N. Y., where was a fruitful field for his zeal and activity. It appears that he found this a most useful and agreeable experience, for he makes frequent reference to this period in his later notes and in his retreat work.

One of a group of charitable ladies, on a visit to the Hospital, had brought a telegram for Father Drum. She did not know him by sight, and on meeting a young priest, dark and dignified in appearance, but very courteous in manner, asked to know if he was Father Drum. "Yes," he answered, "I am the Drum, but not the whole band." It was this pleasant mode of address that often won him many friends.

At the beginning of September, 1906, passage was booked for him on the Cunard Liner *Ivernia,* sailing from Boston for Queenstown and Liverpool. He made all necessary arrangements for saying Mass on board every day when the weather would permit, having previously provided himself with all the required faculties and procured a portable altar out of the funds which, with permission, his mother had given him. He was fortunately able to say Mass every day on the passage across the Atlantic. As time went on, there were to be frequent references in the diary and in his letters home, all manifesting the same devotedness and constant fidelity to daily Mass, no matter what the difficulties of the journeys he had to make.

His brother Hugh once related to the writer how deeply he was impressed with the devotion that Father Drum always showed for his daily Mass. "Walter often told me that since his ordination he had never missed saying his daily Mass; that even in Syria and in difficult journeys around Jerusalem, he was able to have this wonderful privilege." A paragraph from the diary, written later, on the way to Syria, gives us Father Drum's own sentiments on this subject. Shortly after boarding the Messageries steamer at Marseilles he had fallen dreadfully seasick. Yet he writes:

> Oct. 12. I rose at 5 a. m. The sea was too rough. My stomach heaved and gave way again. I had to retreat. Tried to rise again at 7.30 a. m. I felt better, arose and went on deck. I was weak. There seemed no possibility of my saying Mass. I had vomited as late as 9 a. m. The risk was too great. I prayed and prayed the dear Lord to grant me strength to say Mass. You know, Mass is almost everything to me in the day of toil or of pleasure. All else is incidental to the great sacrifice. The Lord was good. By dint of hard work and will-power, I said Mass at 10 a. m., and was, oh, so happy.

In a sermon which he had preached in the Carmelite Church in New York, on the preceding Easter Sunday, there was a long description of the sad scenes of Holy Week, by contrast with the joys of Easter day. A paragraph from this sermon sums up his devotion to the Holy Sacrifice:

> So too was the priest sad and lonely. I do not think you can more than fancy that sadness and loneliness of the priest, during those three days. The priest lives by his Mass; in the strength of that morning sacrifice he goes to the work of the day, and be that work hard, saddening, discouraging, heart-breaking, sickening, he looks forward ever to the joy and comfort of the morrow. And the joy and comfort of the morrow always come, when he brings the dear Lord down to God's altar and lifts him up for the people to adore.

All through his career this devotion and reverence for the Mass may be said to have been one of his most notable traits. Many letters written since his death recall this same fact. One correspondent refers to "his blessing at the end of Mass, that was given with so much reverence and so much dignity in every movement, that I always felt it came straight from the heart of one who lived in God's presence, and was closely united to Christ."

There were several clergymen of various denominations also on the list of passengers on the *Ivernia*. With these and others he at once made himself pleasantly acquainted. It was his practise to adopt a friendly manner with people, on his travels, and would allow them to ask questions, and in this way was often able to do a little good work, especially with people who perhaps had never before come in close contact with a priest. After the first two days, Father Drum was selected by a general vote to be the "Reader" at all the religious services aboard the vessel, even the Protestant clergymen giving him, willingly enough, their own votes for the position.

The voyage itself was rather uneventful, and on September 11 the *Ivernia* arrived at Queenstown. Father Drum arranged his itinerary so as to have about ten days in Ireland. He intended to use this interval to visit his father's relatives in the neighborhood of Killeshandra, in County Cavan; and if possible, to see something of the scenery of the land he loved with such a religious passionateness. A cousin of his, Patrick J. Reynolds, thus describes his recollections of this visit of Father Drum:

> He came to Killeshandra, County Cavan, to visit Mrs. John Reynolds, his father's youngest sister (whom he always addressed as Aunt Sarah). It was characteristic of him to be keenly interested in the places and scenes which his father had so often described to him. He visited and took photographs of the old house of the Drum family, where his father was reared and lived the first fourteen years of his life, and the farm where he worked as a boy. He went through every room in the house his father was reared in, and blessed each of them. The little churchyard where the graves of his grandparents were situated amid such peaceable surroundings, also greatly interested him.
>
> To his great regret the short time at his disposal (he arrived on Monday evening and left at noon on Wednesday), did not permit him to visit his father's relatives who lived at a distance from Killeshandra. The same cause also prevented him from visiting historic Cloughtoughter Castle and beautiful Killykeen which his father often described as the scenes of gatherings in his time.
>
> On reaching Rome he had a private audience with Pope Pius X and it shows his thoughtfulness for his father's

people, that he had a photo-card of the Pope specially blessed by him for Aunt Sarah and the family.

During his studies in Syria and Austria he constantly corresponded with his aunt in Killeshandra and his letters were always cheering. He gave us descriptions of life in the countries he passed through. When sickness or trouble came to Killeshandra relatives he did his best spiritually to encourage and console them.

Every phase of the Irish struggle for freedom was followed by him with keen interest, and, as expressed in one of his letters after the 1916 rising, he had the Fenian spirit of his father in him. He was puzzled for some time by the Irish attitude toward the world war as outlined by De Valera, but later came to see that it was the proper attitude for the leader of a small nation struggling to be free.

He seemed delighted, each St. Patrick's Day, to get the sprigs of Shamrock from his aunt in Ireland, and often related in his letters how in Austria and in Syria and also in America he distributed them among his Irish friends.

He spoke of a wish to visit Ireland and in his last letter seemed in good spirits although working hard at his various pressing duties.

The older people who made his acquaintance were impressed with his religious fervor and the love he had for everything Irish.

Fortunately for the purpose of this biography, Father Drum began from his first landing in Ireland to jot down his daily experiences. The journal is often carefully written, and reveals an alert eye for the beautiful, the religious, and the artistic in the historic scenes through which he passed. The following extracts will show his method, and here and there reveal how thoroughly he enjoyed and appreciated the loveliness of Ireland's scenery, and also, what he looked for chiefly in all his travels. One may readily see that what he wrote down was all intended for future use in sermons and retreat work, and that he was keenly anxious to retain all the vivid impressions he gained on his travels:

12 Sept. '06,
Roche's Hotel Glengarriff.

Yesterday at 8.30 p.m. we left the *Ivernia* and took a tug for Queenstown. The run of half an hour was rough. I took a 9.15 p.m. train for Cork. Went 3d class. On the trip we had a jig by an Irish lassie and a concert by a fiddler. By 10 we were at the Imperial Hotel. I made arrangements to say Mass at the Church of Fr. Matthew, the great temperance advocate.

In the morning the Capuchins were most cordial with me and gave me breakfast. They told me our new General is Father Wernz.

Went out to Blarney Castle and kissed the stone. Heard the bells of Shandon. Saw Queen's College and several Catholic churches.

13 Sept.

The scenery in Glengarriff is lovely past the telling. For twenty miles we coached to Kenmare. The road was gradual ascent and. was very circuitous. The glen is called *garriff*, "rugged," from its striking rough and severe appearance. To me it was not rough and rugged. Our bare and barren cliffs in the Rockies have a ruggedness and a severity that is not seen hereabout. The scene in Glengarriff was soft and sad. Off in the distance were far away cliffs of which not a stone was barren. The rocks were clothed with moss and heather. Here and there the rich and black peat bed took the place of the rocks. The outline of the waving peaks, as they stood over against the Irish blue sky was grand, but the grandeur of the glen was of a sort with the grandeur of the sadness of Ireland. I was charmed and grieved by the scene. The day was clear. The sky had only clouds enough to cast upon the sunlit cliffs and hills a shadow that here and there darkened the green of the moss and the purple of the heather. Not a rock was bare. No trees marred the simple softness of the hills. The Irish heather and Irish furze and Irish moss and the Irish green of the rich grass made a loveliness I had never yet set eyes upon. I looked and thought. I thought of the days when Irish Kings ruled over a land of joy and plenty—the joy of faith and the plenty of the produce of the land. I thought of the churches these kings raised up and England took from us. I thought of the faith St. Patrick brought to us and England failed to ruin. I thought of the land lying fallow because of the laws that made it not worth the farmer's while to work the land. My thoughts might have been rather sad, did I not think of the faith that came to the land of my birth by the Irish bishops and Irish priests and Irish unskilled laborers and servant girls —heroes all, heroes of the poverty of the cross of Christ. We wound along a circuitous route that showed us one side of the glen and then the other.

The fuchsias at Kenmare were marvellous to me. They

make a gorgeous hedge and become trees ten or twelve feet high.

From Kenmare to Killarney we journeyed twenty-one miles. The coach and four made first-rate time.

The journey from Kenmare was more down than up grade. This part of our trip was even more pleasureable than the first part. The softness and sadness of our route was now even more marked than it had been before we reached Kenmare. There were no trees to mar the view. The holly had seemed never so green. The Irish heather had seemed never so deeply purple. Oh! the lakes and streams and mountains and waters and dells and heather and peat and wild growth of Killarney were beautiful beyond the telling. We passed Lough Looscannagh, a deep lake at the end of the glen, and there saw the first of the three famed lakes of Killarney.

Our road took up to the east of the lakes. We skirted the water's edge much of the way. We followed the long range, which joins the upper to the middle (or Muckaross) Lake. High at our left was the Eagle's Nest, not so moss grown and green as old Torc which lifted its giant form at our right. The glory of the day added to the glory of the scene.

14 Sept.

This morning I said Mass at Loretto Convent. I had left the coach last night, when we were a couple of miles from Killarney, and had called on Mother Dosithea, the Superior of the Convent, and a first cousin of my mother. She was delighted to have me say Mass in her convent. My hotel, the Sheheree House, is near Loretto and three miles from Killarney. Sheheree means "forts of the fairies."

About 11.30 we started on ponies thro' the Gap of Dunloe. We rode about five miles through the most picturesque series of valleys I have ever seen. Before we got our lunch we were drenched with rain. The boat ride down the fourteen miles of lakes and rivers was enjoyable in spite of the rain that now and then poured down upon us.

15 Sept.

Mass at Loretto. Good-bye to all. The Sisters sent me to Muckross Abbey in their jaunting-car. This old abbey is one of the best specimens I have seen of Moorish Gothic.

I changed at Mallow and Charleville, and reached Limerick at 3.30 p.m. The Jesuits at Crescent House were busy chiefly in the confessional. I went out to see the town and hired a car to take me to Mungret at 6.15 p.m. Spent most of the time in St. Mary's beautiful cathedral. The altar stone eleven inches thick, lies aside useless. The hard oak *misereres* are still preserved, but not used by the Anglican divines. I enjoyed fully this grand old church of pointed Gothic with its turrets and tower of Moorish style.

Mungret College. This Jesuit college is made of several splendid buildings. The Irish sandstone has a very firm appearance. The Rector of Mungret showed me everything. The boys here stand very well in Irish university examinations. By the side of the college are the ruins of old Mungret Abbey.

16 Sept.

At 4 p.m. I started for Killaloe; about twelve miles off. My train (the only one by which to catch the Monday boat) must have been an express; it covered the twelve miles in forty-five minutes. Some trains take an hour and a quarter to make the dash from Limerick to Killaloe.

Killaloe is built high above the left bank of the Shannon. The quaint walls that girded the thatched shacks of the town seemed to me far more interesting than useful. There is an old sixth century chapel here, the cell of St. Lua. It is the very best preserved of the cells of the old Irish monks. All the work of the roof and walls is of stone. Near this beautiful old cell is St. Flamman's cathedral, built in the twelfth century by O'Brien, king of Thomond. Herein lie the remains of the founder of the cathedral and of a son of Brian Boru. I enjoyed the beautiful old cathedral of Killaloe, though my heart beat with indignant emotion as I listened to the Anglican service and noticed the six girls of imported stock who made up the congregation.

The hours I spent in Killaloe were delightful in historic tours. Everything was within easy reach. I walked to Kincora, the home of our great Brian, and saw the ruins of his fort.

My first duty had been to arrange to say Mass at the Convent of Mercy. The Sisters were delighted to meet a Jesuit.

18 Sept.

I said Mass in the home of my aunt. The customs of the days of persecution still hold here. The two priests of the town live apart and say Mass in their homes. Any house may be used for Mass.

19 Sept.

Mass at my aunt's. Took train at 11 a.m. Stopped an hour at Cavan and saw Cavan College, the Episcopal college for the Catholics.

I spent thirty-five minutes in Drogheda, and by means of a jaunting car saw the whole town. I saw the fine old St. Mary's Abbey and its ruins. The English drove a street through the main door, up the broad aisle and through the rear wall. The left wall is part of a saw mill. Drogheda was the first town against which Cromwell directed his cannon. The old town walls, the tower of St. Lawrence, Magdalene tower and other relics of the old Irish chiefs were of great interest to me.

21 Sept.

At 1 p. m. I arrived, after a slow train-ride at Maynooth. The lackey at the door was in uniform. There is much more of ceremony in Ireland than in the States. The porter of Gardiner Street residence, for instance, wears a uniform. The Rector, Mgr. Mannix, treated me with wonderful courtesy, introduced me to Dr. Beecher (who had taught in Dunwoodie Seminary, N. Y.) and to Dr. Boylan (who had studied Oriental languages in Berlin), and insisted on my staying for dinner. We chatted a very great deal. I found the faculty of Maynooth to be splendid priests, learned and impressive. Maynooth is by far the most wonderful seminary in the world, and the chapel is the most magnificent college chapel I have seen. It is pure Gothic, and cost half a million. Maynooth has taken up the Irish language with zest, and will carry the use of the tongue of our ancestors far and wide throughout Ireland. At 5.30, Mgr. Mannix had me driven to Clongowes in his brougham.

Left Clongowes and had to walk to Sallins, a distance of about five miles, along a dark and dreary road. It was my thirty-sixth birthday.

22 Sept.

At 11 a.m. I started on the *Scotia* for Holyhead. Dublin had interested me somewhat, but not so much as had the less modern part of Ireland. St. Patrick's Cathedral and Christ Church are the very best specimens I saw of the old Moorish Gothic, which is so common in Ireland. Both churches, of course, were borrowed from us by Cromwell and never returned. The Vicar-Regal Lodge was too modern and not grand enough to interest me. Other city and government buildings are such as you would find everywhere; those of Dublin are Corinthian and Doric.

At 3.30 p.m. we left Holyhead. An Englishman pointed out to me Snowdon, the sands of Dee, the fashionable resort, Llandudno and Conway Castle. The grandeur of Killarney and its loveliness cannot be set by the side of anything of this Welsh country. I left the train and the express route so as to see the pre-Reformation cathedral of Chester. Here the old walls of the Roman town are still intact. The cathedral is the grandest I have seen. The grandeur of the naked sandstone appeals to me. I walked through the cloisters and from its fine old passes heard the excellent polyphonic music of the male choir.

At 10.30 p.m. I reached Birmingham, tired and sleepy.

23 Sept.

I said Mass at St. Chad's, Birmingham, where Canon Keating treated me most courteously. St. Chad's is a post-Reformation church of grand pillars and an exceedingly high arcade. The exterior is a poor affair of brick. The interior would be splendid had it not been spoiled by bad taste. A gaudy and glaring stencil fresco brands the interior as inartistic.

24 Sept.

I said Mass at our church of St. Aloysius, Oxford. This beautiful Gothic church has been ruined by paint.

After breakfast with the Fathers, I went to Pope's Hall. Only Fr. Pope was in.

I went through all the buildings thoroughly. I went over the ground hallowed by Campion, saw the library presented by De Bury, walked over Newman's walks at Oriel, and gazed upon the once Catholic halls of old Oxford.

25 Sept.

I saw no sights save from the top of the bus. I am tired of sight seeing and in need of rest. I shall see London on my return. Of course I saw Piccadilly, Regent St. and the outside of the great churches. Our church in Farm St. is wretched in outside appearance but beautiful inside.

At Farm Street he had the good fortune to meet and consult with Father Strassmeier, the great authority on cuneiform inscriptions, and also Father Deimel, the Assyriologist. The next point of interest in the diary is from Paris:

28 Sept.

Mass at St. Roch's. I went to Cook's and found there would be no room for me on Messageries steamers till

4th of Oct. I then started to look for Jesuits and found them on Rue Buonaparte, Rue du Regard, and Rue des Vieux Colombier. A scholastic took me out. I saw the Louvre and Tuilleries. I saw many Jesuits. They live in apartment houses and in small groups of two and three. They meet for community exercises. About twenty of us dined together at Rue du Regard. To house me the Fathers put out their servant from the rooms on Rue des Vieux Colombier. We were three in the suite; Fr. de Fraguière and Fr. Prat, a member of the Biblical Commission, and I . . . It was delightful to talk with most simple men and then to find that they were the great Billot, professor of dogma in the Gregorian University, Rome, or the pious and learned young Father Lebreton, prof. of dogma in the Institute of France, or other men of fame.

1 Oct.

Mass at Montmartre at altar of St. Ignatius. Breakfast with Jesuits nearby. I then saw the Basilica from top to bottom. It is grand. I never thought I should like Roman Byzantine as well as Gothic. The grandeur and splendor of Montmartre have changed my ideas.

We went to St. Peter's, a thirteenth century church near Montmartre and to the chapel of St. Denis, where the Society was formed. Then visited our college of St. Ignatius. It had eight hundred pupils and now has two hundred. Our Fathers keep away from it.

In the afternoon a Jesuit took me to our Church at Rue des Sevres. What a touching scene. I wished to kneel at the tomb of FF. Olivaint, Clerc, de Begny, and their two companions who were martyred during the Commune of 1871. The *concierge* would not let us into the Church. The government had taken complete possession. She was under suspicion, and was going to be sent away on the morrow, though she had been *concierge* here for twenty years. We urged that if she be sent away, she might as well run the risk of letting us in. We offered her two francs. She could not be bribed. But at last she began to suspect who we were, and opened the church door. What a sight met us! All the side altars were walled in with planks, the sanctuary was turned into a stage and a concert hall was made of our church. We went behind the planks to the side chapel where our dear relics lie, knelt down and prayed, and kissed the stone that marks the tomb. *Est-ce que vous êtes des pères?* the poor woman cried. We said we were. She wailed and moaned. I went away sad, yet strong—strong in the strength of my vocation.

2 Oct.

Mass at Jesuit chapel, Rue du Regard.

Father Minister took me about in the afternoon. I first saw Ste. Chapelle, a little gem, in purest Gothic, which has converted me to *some* love for fresco of stone. It has the most delicate and dainty and artistic fresco. The chiaroscuro is splendid. The old stained glass of six centuries ago allows the light to enter into the chapel in lovely tints.

We then saw the restored palace of St. Louis IX. Here is the famous *Court de Cassation*, where Dreyfus was made notorious. The modern frescoes and decoration compare favorably with the wonderful work of Versailles.

The most interesting part of the day was a visit to the Rue Haxo, where our five fathers were shot during the Commune.

3 Oct.

Mass for the family intentions, at the famous shrine of Notre Dame des Victoires. I had arranged several days before, and had chosen the first Mass, that of 6 a.m., with the hope of starting before the time for the sake of greater devotion. Though I reached the church at 5.30, it was not opened till 5.50. What a disappointment! If ever I wish to say Mass slowly, it is when I am saying a Mass of special devotion. The half hour was passed by three minutes, and my successor was waiting at the foot of the altar, when I shyly made my departure. If I have anything to do with shrines, I will not have the hour of the Mass fixed with mathematical accuracy. Why, it spoils my devotion, when I have to hurry. Well, never mind; I prayed much for you all [this diary was written for his family] during my Mass and during an hour or so thereafter.

6 Oct., Saturday.

I was on the way to Paray-le-Monial, arriving at 5.50 a.m. Mass for family intentions at the Altar of Apparitions, at 6.30 a.m. I had intended to finish Mass in half an hour, but failed. I was so happy and filled with consolation that I could not keep my emotions under control. You may be sure I prayed much for you all to the Sacred Heart, as I stood up there at the altar of our dear Lord's apparition and offered up the great Sacrifice to Him whom men so often offend and insult. I never before was in a place of so much devotion, never experienced so great joy as when I raised in my hands Him whom in reparation for my sins I had brought down to that dear altar of the Sacred Heart of Jesus. The remains of Blessed Margaret Mary are entombed in the sanctuary. The very same grille was at my right, and the nuns were behind the grille as Blessed Margaret Mary had been when she received from the dear Lord the message of love and reparation. I offered Him and myself in reparation for my sins, for all the family, and for all men. And I felt, as I never had felt before, that the dear Lord was pleased with my offering. I shall never forget my Mass at the altar of apparition in the dear little chapel of Paray.

After Mass, I saw our Tertianship, now a school taught by seculars, in which Father Pardow was trained. I kissed the tomb of dear Father de la Colombière.

7 Oct.

Mass at St. Sernin, Toulouse. In this church rests the head of St. Thomas Aquinas. At 11 a.m. I started for Lourdes and arrived at about 2. I engaged a room at the Terminus Hotel. The faith and devotion of the pilgrims of Lourdes is remarkable. All scepticism oozed out of me. The enthusiasm of the place possessed me. I am not much in favor of outward signs of faith that are out of the ordinary; yet Lourdes has so much of that outward manner of devotion, that I scrupled not to follow the example of a little Carmelite nun, and to stretch out my arms in the form of a cross in quiet prayer. At Paray, all was quiet, and there were no outward signs of emotion. One felt it was there, but the devotion was Jesuit devotion, quiet and unassuming, a union of the soul of God that was out of the ordinary and yet did not show itself. At Lourdes all was otherwise. Again and again I kissed the rocks on which Our Lady stood, and the ground on which Bernadette knelt; and was most happy in my many prayers for you all.

9 Oct.

I reached Barcelona at about 9 a.m. The customs official wished to charge duty on my altar until he learned I was a priest.

Mass at our church. Rested and saw college. Dinner at noon. Recreation with Fathers. I have been much pleased with the cordial reception I have everywhere received from my fellow Jesuits. They have treated me as one of their own. At 4.30 I visited Santa Cueva, where St. Ignatius fasted and prayed and did heroic penance and wrote the "Spiritual Exercises." The cave is formed by a jutting ledge such as we used to find in the Rockies. How lovingly and reverently I kissed the very rock-bed floor on which our holy Father knelt as he wrote, and the two crosses he carved in a rock of his cell. I prayed much for each of you by name. I visited the very spot on which his poor, shattered and famished body lay during his ecstacy. My joy was full this morning.

10 Oct.

I said Mass in the Santa Cueva. After Mass I saw the Chapel called Rapto, in which St. Ignatius lay wrapt in prayer for days. I saw all the spots of devotion.

At 8.08 a.m. I started for Barcelona. I took train from Monistrol up the mountain, saw Montserrat and the statue before which St. Ignatius kept his vigil of arms. I left Monistrol at twelve arrived in Barcelona at 2 p.m., saw the University of Barcelona and much of the city, and started for Marseilles at 6.31 p.m.

The following note was written just before sailing time on board the Messageries steamer from Marseilles to Beirut:

Paquebot le Saghalien
11 Oct. '06.

My Dear John,

I am about to start for the far-off land. Pray much for me and have the family to pray.

My roommate bids fair to be decent. He seems to be an Englishman. I have heard him say only a word or two.

Today there has been more rain than I ever before experienced out-of-doors. I had to stay out, in order to do business. The rain fell in never-ending torrents. I failed absolutely either to find the Jesuits or to trace letters. You know, the catalogue gave the address 62 Rue de Grignan. I went to that address and found no one. All my inquiries were unsuccessful.

I must hurry to have this ready for mail. God bless and care for you all.

Lovingly,
Walter.

Let us go once again to Father Drum's diary and follow him in his way to Beirut.

13 Oct.

Mass at 7 a.m. and at 9 a.m. I went ashore in the launch and saw Naples. Had six hours to see the city. Saw the churches. Our Gesu is a splendid structure in Roman Byzantine. The remains of St. Francis Jerome are in this church. The cathedral has a chapel built in the fourth century. Also saw the Museum. We left Naples at 1 p.m.

14 Oct.

Mass in my cabin at 9 a.m. My roomate did not rise in time. We are about fourteen priests on board and fifteen or twenty ministers and so-called missionaries.

15 Oct.

Mass at 9.30 a.m. We stopped at Athens, or rather Piraios, from 3 to 9 a.m. Dr. Fortescue, a Frenchman, named Fontaine, and I saw Athens together. We took a cab to Athens, and by means of a guide saw the whole of the ancient city—the Acropolis, prison of Socrates, Parthenon, Temple of Zeus, Temple of Erechtheus, Areopagus, Museum, Temple of Apollon, etc., etc. The day most interesting; though the guide talked too much nonsense. We partook of a Greek dinner and found it rather modern in its details. The new stadium is splendid. It cost 8,000,000 fr., a fabulous sum for the poor Greeks.

16 Oct.

Mass at 6.25 a.m. Fr. Fortescue is now my roommate and served my Mass. We are now running up the Gulf of Smyrna.

It seems strange to see Mytelene and Chios and other islands of which I read so much, and to find them mostly wild and barren and rugged.

The Ionian Sea and its many isles of Greece are lovely, and the Ionian sky is as blue as the Ionian waves.

I went ashore to see Smyrna with Dr. Fontaine, of the Treasury Department of France, who has been most cordial with me. We arranged with a guide for passage to and from the ship and for service. We called on the Lazarists who occupy the property once held by the old Society. The Jesuits have not done work in Smyrna since the Society's disintegration under Pope Clement XIV. We called on a Silesian brother, who insisted on showing us the modern town. We passed a Turkish College. The Director showed us every part of this college of more than five hundred students.

17 Oct.

I arose bright and early to note the entrance to Hellespont. The mist was disappointing. I was sick from jolting and went back to bed; but later was in fine trim for Mass.

We are now in the Sea of Marmora and are pitching against head winds.

18 Oct.

Last evening we entered Stamboul before sunset. The sight was most beautiful; on our left, the ancient Byzantine separated from Pera by the Golden Horn, on our right the Turkish addition to Stamboul. As night came down, the minarets of the various mosques were lighted. The city is built on many hills and its lights were beautifully laid out. I tried to disembark, but failed. The Turkish official said *Demain*.

I was sick most of the night, but at 8.10 a.m. said Mass and later tried to leave the ship, but the passport official would not let me by. My passport was incorrectly visaed. I was one of about twenty who could not leave the ship. I sent a letter by a guide to the Jesuits; they arrived at 11.30 a.m. and spent half an hour trying to free me. It was unavailing. I spent the day aboard the ship. The Fathers came back to see me at 3.30 p.m. We left Stamboul at 4 p.m. We are now in Hellespont.

19 Oct.

We kept between Mytelene and the mainland. At 12 we reached Smyrna. I went to Ephesus with Cook's party. A special train was chartered. Ponies awaited us at Ephesus. We rode out to the ruins. The new town is about three miles from the chief part of the ancient city. I saw the theater where Paul spoke and the forum wherein John gave his lessons of love. Then too, there was the great basilica in which the Third Council, that of Ephesus, was held. Here St. Cyril thundered forth the dogma of the Divine maternity of Mary.

21 Oct.

I said Mass on deck for the passengers. This time I took the initiative and had two of the important ladies to protest against the loss of Mass. The Captain yielded and had flags and flaps arranged to form an improvised chapel. Some of my Protestant friends were present.

22 Oct.

We reached the harbor at 5.30 a.m. The Fathers came for me and here I am in the Jesuit University of Beirut.

From Beirut, Father Drum sent to Father Frisbee, then editor of the *Woodstock Letters*, eight or ten articles of considerable length, dealing with various subjects of interest in the Orient. In one of these articles, he gives this description of the approach to the city:

> As we steam towards Beirut, a striking panorama lies before us. There are no islands, no bending capes, nor any harbor to obstruct the view of the city. To the left rise the Lebanon in rugged relief, clearcut as are the mountains of Killarney. They seem near because of the clear Syrian sky, and yet they are far away. Behind the Lebanon towers Sannin, snow-capped all the year round. In front of us is Beirut, a city set upon a hill. What a splendid sight it is, resting glorious upon the jutting headland of Ras Beirut, that forms the bay of St. George! The staid Fr. Barnabé, in his new "Guide to the Holy Land," borrows an Arabic figure, and says Beirut looks like a lovely Sultana reclining upon a couch of verdure and gazing in an idle day-dream at the waves that roll ashore.

One of the passengers on board ship from Constantinople to Beirut was the Princess Gagarin, an able and highly cultured lady, who was the wife of the Russian consul-general at Beirut, and through him related to the famous Madame Swetchine.

> We had an extended conversation on the Bible and the early Church. She is remarkably well read, and speaks English beautifully. I had several talks later. She tells me people are afraid of Jesuits; because they are so well-educated, and talk so well that they make the worse appear the better side. She is a schismatic of the Russian Church. I had another passenger, a Madame Flament, to talk to the purser about her right to hear my Mass on board the *Saghalien*. The woman won where the priest had failed. The Princess was greatly impressed. She admired the work of the Jesuits in Syria and the books published by them in Beirut. All her books were from the Jesuit press and bindery in our college there.

There were the usual passport and customs formalities to be gone through, and although there was little difficulty there, because the Fathers had made arrangements for him beforehand, the diaries make humorous record of the vexatious delays he encountered at other times in passing through the Turkish customs, especially at Constantinople. He was amused to find that his own first name and his father's first name were all that were essential for a passport. He was henceforth to be, officially *Abouna Waltar ibn Johann,* (Father Walter son of John).

As regards the city itself, he was favorably impressed for the most part, and was especially interested in the historical and Christian associations of the city. The famous *stelae* on the Nahr el Kelb, in the neighborhood of Beirut, claimed an early visit. There was the church of the famous Bleeding Crucifix, and the tradition of St. George and the Dragon—these and many other subjects of interest are vividly described. In one place mention is made of a distinguished American who was entertained at the college a few weeks after Father Drum's arrival in Beirut.

> The city's great pleasure grounds are the famous Pine Groves to the south, where a race-course forms the chief attraction. Tourists are always brought to "The Pines" as they are called, by the dragomans, or guides, who spin out yarns about the famous grove. The Honorable W. Bourke Cockran was here lately, and laughingly told me his dragoman had pointed out the pines and tried to convince him they were the Cedars of Lebanon.

Mr. B. Cockran wrote him before returning to America:

> A thousand thanks, my dear Father Drum, for the "Sainte Bible" which you have been good enough to send me. It will add much to our glimpse of the Holy Land and serve to remind us of how much I owe you for a pleasant recollection of Beirut.

It was found that classes were not to begin till November 3, and accordingly the intervening time was utilized for a trip to the Lebanon. A letter to his brother describes this expedition:

> Damascus, Syria.
> 29 Oct. '06.
>
> My dear John,
>
> Be patient, and in course of time, my diary will be set in order, and you will have a full account of my trip to Beirut. I arrived in Beirut last Monday. By the way, either Beirut or Beyrouth will do. The Arabic is neither. Your two letters of ———, well, I don't remember the dates, came in due time. I fear that many of my postcards to you may not have reached you. I sent them by Turkish post, so as to send you the Turkish stamps. Just before reaching Beirut, I learned that Turkish post was not safe. At Beirut, you know, or at most Oriental ports, there are post offices and postal service for each of the great European powers.

I am in first-rate health. Please see to it that my letters go the rounds of the family, and spare me the time it would take to write up the same things for each. Later I shall write to each. Our courses do not begin till Nov. 5. Five of us Orientalists are on a tour with the Professor of History of Syria. We started at 7 a.m. the 27th inst. Beirut is only about sixty miles from Damascus; but half of the route is up an incline. · We covered the sixty miles in thirteen and a half hours, and reached our residence here at 8.30 p.m. The train moved very slowly up the incline and was delayed for three or four hours some twenty miles from here. There had been a derailment of a train the night before; we had to walk some distance and be taken up by another train. We had a superb view of the Lebanon and Anti-Lebanon. The wild scenery shows no traces of trees in some parts, and the hills stand out against the clear sky with a boldness like to that of the hills of Killarney. The Anti-Lebanon hills are chiefly calcareous. Their steel gray coloring was beautiful as we viewed them under the light of the full moon.

On the 28th, we went through the bazaars and saw very thoroughly the Oriental life. Damascus is so far inland as to be more truly Oriental than are most of the great coast cities. The costumes of the people and their customs are most pronounced, branded with the Oriental mark.

We saw the road by which St. Paul entered the city. The traditions about St. Paul here are few and doubtful. We saw that part of the old city wall, wherefrom the Apostle was let down in a basket; the house that is said to have been that of Ananias, and to have been used by St. Paul for the spread of the faith.

The grand old mosque, once a pagan temple, then the Church of St. John, was the most beautiful basilica I have seen. Its length is enormous, the Corinthian columns are splendidly wrought, the modern work of inlaying of mother of pearl and of mosaic is exquisite and most delicate.

Whatsoever we saw of Damascus has given me a truer idea than I could otherwise have formed of Oriental life just as it is. The Turk sometimes makes a vow to feed the dogs once a day. I saw a Turk feeding a pack of snarling, wolfish curs with a large basketful of bread. Dogs are most numerous in the Orient, and look rather vicious. They seem harmless, and are undoubtedly privileged creatures. We saw the Syrian leper colony, whose horrors you can fancy.

As we returned from the part of the wall, through which St. Paul made good his escape, we met a Greek Catholic funeral cortège. Officials led the marching column. Then came torch-bearers and many priests chanting a most lugubrious whine. The corpse was in a beautiful white coffin, which was borne by men. There were no carriages.

The finest Catholic church is that of the Greek Catholics whose Patriarch lives here. I was surprised to find a special place for women to receive Holy Communion, in the rear of the church. Only men receive in front of the altar.

Today we visited the cemetery, and saw the tombs of the great Pashas and other "swells," went out to the suburbs to get a view of the plain of Damascus. What most interested me today was a visit to Nasson's manufactory of Oriental goods. The wonderful works in brass and gold and silver were most interesting and cheap.

God bless you. Pray for me. Tomorrow we go to Baalbek, the ancient Hieropolis. We return to Beirut on the 31st.

Love to all.

Lovingly,

Walter.

In the articles sent to the *Woodstock Letters,* Father Drum speaks favorably of the even climate and healthy air of Beirut, despite the greater heat that prevails there; but records that he was frequently repelled by the lack of sanitation and of paving throughout the city, and of the iniquities of Turkish misrule in Syria. His diary, which took the form of a letter that was filled out as he went along in his travels, ends with the arrival at Beirut, and was not taken up again till the following spring, when he was again traveling, to Palestine. This first part concludes as follows:

I am putting on the customs of Beirut as quickly as I can. There is nothing like adapting one's self to one's surroundings. Such action makes for charity, and charity is the great virtue that makes the Jesuit life happy.

I am very, very happy. Our studies began yesterday, Nov. 3. I shall work at Arabic and Syriac. My health is splendid. The climate here is delightful. We have tropical weather, but the heat is not now oppressive. We are clothed to suit the climate. My white stockings, and close-cut and tonsured head, and bristling beard would look fierce in the States. They are just the thing for Syria.

Later on I shall write to each of the family. I hope my lengthy descriptions will be read with pleasure by all. My schedule varied as I went along. I heard in Dublin that there was no hurry. In London, the Jesuits told me that the Professor of Hebrew had not left England for Beirut. Hence I took my time and traveled according to the desire of the day. Had I fixed a schedule and rushed to fill it out, I should not have enjoyed my trip. The long journey, as you will have observed, has brought me much good and pleasure.

I hope there will be no need of intervention in Cuba. The moral effect of the presence of five thousand soldiers will be good. I wish you would each of you send me clippings of things you know will interest me. . .

This morning I received a visit from a Mr. Bassone, a Syrian, a graduate of Beirut, who was on the *Saghalien* with me and became very friendly towards the end of our voyage.

My dear Joe, no word from you! Do write! I pray much for you. You spoke of some gift I should receive from B. Maybe it went to B.C. and stayed there. I saw none. Give my very best wishes to B. and wife. God bless your work.

My dear mother, thank you very much for your love and kindness. Be sure I pray for you often in every day. Have no fear. I am well and with those that love me. The Fathers are most kind. I will write you about our school later on.

God bless and love you all. Remember that, though two oceans keep me from you, my heart and prayers are with you, mother, John, Mary, Joe, Al and his wife and boy, Hugh and his wife.

Lovingly,

Walter.

CHAPTER IX

AT BEIRUT

With regard to the purpose of his coming to Beirut, and the plan of studies that he was to follow, it appears to have been the original intention of Father Drum to study and travel for two years in the Orient, and to spend a third year in some German university. His subjects, as we have seen, were to be Oriental languages and Holy Scripture. He had, therefore, to lay a thorough foundation for future work, by his studies in Beirut. The purpose of the third year was to acquaint himself with the methods and the teachings of present-day rationalists at their very source, in order to be able to meet them effectively and on their own ground. We shall see in the sequel that this plan was never carried out in its entirety, and the last part especially was cancelled by Superiors as being unnecessary and dangerous.

When, however, Father Drum found himself settled down at Beirut, he plunged at once into the serious study of Arabic and Syriac, together with Biblical archeology and such kindred subjects. In spite of his years—he was thirty-six when he came to Beirut—his unusual gift for languages and his well-trained memory gave him an exceptional advantage in these studies. The university provided excellent and thorough courses in all subjects that pertained to the languages of the East and to the study of the Bible, and for such advanced students as Father Drum a definite program of studies was available under a specially appointed Professor. Besides, there is a splendidly equipped library at the disposal of Orientalists and Biblical students. The university also provided opportunities for scientific excursions throughout Syria, Palestine, and Egypt, the expenses of which were defrayed in part by the French Colonial Government.

It may be expected that a man like Father Drum would not fail to take advantage of all these excellent aids to his studies. Indeed, it is hard to see in what detail he failed, there was so much of hard work and travel and exploration that he managed to crowd into his stay of less than a year in the Orient. He availed himself of every possible opportunity to see and learn everything that was worth while in the East, by meeting and talking with the native people, acquiring their difficult language and gaining an insight into the customs, the religious character and the mode of life of the people of the Orient that an ordinary individual, less alert to his opportunities and less well-equipped to take advantage of them, would scarcely acquire in a stay of three or four years in the country. Faithfully he noted down, day by day, his ever increasing store of impressions. The diary makes very interesting reading, although it must be remembered, when we note certain disagreeable features, that his impressions were gained in the Near East as it was nearly twenty-five years ago, and that the conditions he described have undergone considerable changes since that time. Not only the Great War, but the peaceful advances of the last decade or two have brought about an entire transformation in Asia, while the great material improvements of Syria and Palestine have kept pace with the political revolution that has followed the war.

However, Father Drum's observations are interesting for their own sake, and for the invaluable use he could make of them in his retreat work and in his preaching. His mind came to be saturated with Biblical and Oriental color, and he brought away with him the flavor of the East that was to manifest itself in everything that he wrote or spoke. The series of Bible lessons in the *Pilgrim of Our Lady of Martyrs*, from 1913 to 1918, and in the *Queen's Work*, from 1918 to 1920, and all his sermons and retreats abound in imagery and illustrations that show how intimate was the knowledege he acquired of the customs of the Syrian people, their simple strong faith and their difficult language. The following selection, besides describing a beautiful center of pilgrimage in Syria, gives also his estimate of the religious fervor of the people of Mount Lebanon:

A Shrine of the Lebanon

The famous Lebanon is almost entirely Christian. Here and there are a few Druse villages; but the orthodox Mussulman generally stays elsewhere. Among the Christians of the Lebanon, the vast majority are Maronites; and the Maronites are in the East what the Irish are in the West—steadfast Catholics in life and in death. The Maronite is the only Eastern rite that is not sundered by heresy. Even the up-to-date Protestant missionary, with his money and schools and medical supply, is as ineffective today in Lebanon as he has ever been in Ireland. The priest, on the contrary, is the privileged character in the country; his *khouri* is to the Maronite all that the *soggarth aroon* used to be to the Irish. The little ones run after us to kiss our hands or to ask for a picture. The grownup people have a special salutation for the priest only,—*Medgilla,* ("glory be to God"). We answer, *Daiman!* (Forever!) The smiles and joyous faces that greet the priest at every village, make him happy to be a priest, thankful that God has preserved this Catholic people from Moslem fanaticism by the impregnable fastnesses of the glorious Lebanon.

One feature of the Catholicity of the Maronites is their tender devotion to Our Lady. This devotion has led them to erect an heroic statue to our Blessed Mother, thirty feet high, that rests on a lofty pedestal and looks out over the pretty Bay of Junie to tell the passerby that Mary shields the Lebanon from heresy and Islam and infidelity. The same devotion also led to the establishment of a special shrine to Our Lady of Deliverance at Bikfaia.

In the early part of the last century, the Jesuit Father Ryllo returned to Rome after mission work in Mesopotamia. He was a powerfully built and handsome man, and attracted much attention in his Oriental robes. Podesti, a painter of renown, begged the Father to pose for a picture of Solomon in judgment. Father Ryllo yielded on condition that Podesti paint a Madonna for him. To this condition the artist consented.

Father Ryllo sent his precious Madonna to Bikfaia in the Lebanon. The Fathers there, not realizing the value of the picture painted on copperplate about 16 inches by 12 inches, when their chapel was completed, placed above the middle of the altar a painting of St. Francis Regis, to whom they intended to dedicate the building. But the people were not satisfied, and while the picture of the Saint was being set in place they cried out, "Hail to thee, Virgin Mary!" Our Lady was straightway chosen Patroness of the chapel, and her picture found a resting place above the middle of the altar. The good people used to say to the Superior: "Your Madonna is too small for the middle of the altar." His reply was ever the same: "Yes, she is small, but she is mighty." So thought the simple Lebanon folk, and their devotion to Our Lady of Deliverance never lessened.

In 1845, the little chapel of Bikfaia was destroyed by a tempest. As the storm raged Mary's friends prayed. Scarcely had they left the shrine, than it was utterly ruined. The good Maronites deemed they had been saved from death by Our Lady of Deliverance, and set to work at once to erect in her honor a more lasting place of worship. Mgr. Villardell, the Delegate Apostolic, contributed 17,000 piastres; Emir Haidar gave the same amount. A splendid church was built and named Our Lady of Deliverance. It cost only 20,000 francs, but money went a long way in the Lebanon of those days.

Since that time the church of Our Lady of Deliverance has been gradually improved and much enriched. In a side chapel, the walls and windows of which are richly veneered with sculptured marble, the miraculous picture is enthroned upon a handsome marble altar, on which are sculptured Our Lady's symbols, the rose and the lily. The tastefulness of the decorations of this beautiful shrine help much to the devotion of those who honor Our Lady of Bikfaia.

Just outside the house of God, lie the remains of the devoted Emir Haidar. He had always loved Our Lady of Deliverance. During the conquest of Syria, by Ibrahim Pasha, the Emir was exiled to the Soudan and kept there in chains. On his departure from Lebanon for prison someone tauntingly remarked: "We shall see what your Madonna will do for you!" At his homecoming, he gave exultant answer, "You see what my Madonna has done for me." The Emir was, later on, a pious and in every way a just governor of the northern Lebanon.

When the new and massive church was completed, the devotion of Bikfaia, and indeed of all the Lebanon, to Our Lady of Deliverance increased most remarkably. Tapers, jewels, ex-voto offerings, and the ever-present Faithful attested this devotion. Nor was the Blessed Mother forgetful of her clients. In 1860 the Druses began the massacre of the Christians in the Lebanon. Onward they came with fire and sword, and nearby villages were in flames. The armed fanatics were within an hour of Bikfaia. The Christians had no arms, nothing to protect them! Yes, they had Our Lady of Deliverance! To her they prayed. A woman of the family of Emir Haidar, in Oriental abandon, washed the altar steps with tears and dried them with her dishevelled hair. Our

Lady was true to her title and delivered the Christians of Bikfaia. The Druses were terror-stricken and fled in confusion. Later on, when asked why they had fled, they recounted an apparition of a beautiful woman above the church of Our Lady of Deliverance; she ordered them away and they obeyed.

Bikfaia was also delivered from the soldiers of Ibrahim Pasha by the intercession of Our Lady of Deliverance.

The great number of ex-voto offerings, some of which have been sent by the Syrians of the United States, show that even in distant lands they bear in their hearts the love for their Queen. Many miracles are recounted. Father Louis Cheikho, Dean of the Oriental Faculty of the Jesuit University of Beirut, told me that one of the students lay dying at the University from successive tubercular hemorrhages. The doctors had no hope of cure. The good Father, hurrying to Bikfaia, brought back the miraculous image, and applied it to the boy who was completely healed and at once. A paralytic of seventy-six years of age, and a child with typhoid fever, were brought before the picture and straightway cured. The village was once saved from a plague of locusts by prayers offered up to its Madonna.

Many have been the miracles wrought by Our Lady of Deliverance of Bikfaia; but her greatest miracle is the devoted faith of her clients. May she watch over the Maronites in this critical time. Thousands who have gone to the United States, have thought only of gold and forgotten God. Even in the Lebanon a new ferment is spoiling the mass. But the Blessed Mother will watch over them, and through zealous efforts in their behalf deliver them and their children from peril.

Four of the articles which Father Drum sent to Father Frisbee were published in the *Woodstock Letters,* which is a magazine for private circulation only, and were combined under the title "Our Mission in Syria," appearing in successive issues 1907 to 1909. A few selections are here given from some of the other unpublished articles. In one of these, after a detailed description of the customs at a Syrian funeral, he continues:

Such a scene is one of many that an Occidental cannot fairly estimate: it takes years to appreciate the Orient from an Oriental standpoint. Such unrestrained grief is really sincere on the part of relatives, and no comedy. In the Occident, great and noble grief and love are best shown by a suppressed intensity; in the Orient, all is otherwise. A Syrian feels obliged to show his grief. When a child leaves home for a Novitiate, the Syrian Catholic mother is overjoyed; yet she must enact the grief of parting. People would misinterpret a calm, reasonable, and intense sorrow; they demand a show of frantic grief.

In another paper Father Drum gives a vivid account of the troubles, often culminating in massacre, that annually took place in Beirut, in connection with the *Qurban Bairan*, a Moslem festival. He adds a bit of prophetic forecast—this was written in 1906—and the subsequent revolution of 1908, and the events after the Great War, fully proved the accuracy of his summary of affairs at that time:

> It is after such events that one realizes the hopeless condition of native Christians in Syria. They feel they have nought to hope from the Turkish Government. Their hope has been in foreign powers; that hope is now passed. There is in the Lebanon a new generation that will some day be heard from. Unfortunately, the young Lebanon has ideas that are revolutionary not only against the civil government, but against the clergy, especially against the native monks. Time will see a readjustment of conditions in Syria; and it is certain that the new order will be better than the old.

It appears that he made strenuous efforts to acquire the Arabic language thoroughly. Indeed, he told a friend at a later time that the strain to which he put himself nearly resulted in a loss of mind. But as a matter of fact, his progress was so rapid that in December, when a tour was arranged to view the ruins of Egypt, he found himself able to speak the language with sufficient facility to dispense with the troublesome dragoman, whose services are so necessary to the ordinary tourist in the Near East. This trip to Egypt is described in one of his papers at very great length and with painstaking care.

> The French Colonial Government pays our passage [he remarks at the beginning of the account]. Hence the excursions of the Orientalists of Beirut cost very little more than would the vacation spent at home.

There were four priests and three Scholastics in the party that, a week before Christmas, boarded the *Congo*, a Messageries steamer of the line, for Port Said and Cairo. He remarks:

"We tried third-class tickets from Port Said to Cairo." But this act of self-denial proved to be a very trying experience and was not again repeated. He describes his view of the Suez Canal, and of the great moles, and of the statue of De Lesseps at the entrance; and notes the pleasing impressions he received of the modern Cairo, clean and handsome under the efficiency of British rule in Egypt; there follows an interesting description of the visit to the pyramid of Cheops. The party climbed to its summit, and explored the interior under the guidance of an American, named Mr. Covington, who was in charge of the excavations at Gizeh. The ruins of Luxor, Karnack, Thebes, and other cities were visited in turn, and they spent hours roaming among the temples and monuments of old Egypt. Father Drum's descriptions indicate that he was at great pains to examine all the more important antiquities, and at the same time to fit into his own experiences the history and literature which he had read. The details he gives in his notes are far too careful and minute for quotation here.

At one point he pauses to describe a Mass in the Coptic rite which he attended at Mina, in Upper Egypt:

> Most of the Mass was in Arabic, though parts were in Coptic, and stray short sentences now and then were in Greek. In all the Oriental Rites, the use of Arabic brings the people into close contact with the liturgy. In this Mass, there were three epistles chanted by three laymen at the ambo. One layman chanted a great deal, and accompanied his chanting with a triangle which he beat in three-fourth time. The tone of the Coptic chant seemed less sad and more varied than the tone of the Greek, Maronite and Syrian chants that I had heard. A great deal of incense was used, and this with the many different chants made the Coptic Mass inspiring and interesting.

This journey took the party as far as the first cataract of the Nile.

The trip ended on January 4, when the travelers boarded another Messageries steamer for Beirut, arriving there the following day.

A letter to a cousin in Fall River, Mass., describes some details of additional interest in this Egyptian journey:

> Aswan, Egypt,
> Christmas Day, 1906.
>
> My dear ———,
>
> Merry Xmas to the tribe of C———, especially to

Joe. For *you* wrote to me. Thank you very much for that breezy letter of Nov. 10. ......Now, do send on the second batch of kodaks; the first batch is first-rate. I get recreation by looking at myself and the tribe. How changed I am! The whiskers hide me quite.....Write me often—tell me all about your phonics. It is not enough that your thoughts stray to Syria. I call for words, words, words!........

I left Beirut the 18th of Dec. and shall return Jan. 7th. My visit to Egypt is for the study of temples and tombs that abound hereabout. I am most interested in the study of these wonders and shall some day tell you all about them. To visit ancient Thebes and Karnack, we stayed three or four days in Luxor. I rode to Thebes and round about both yesterday and the day before. You should have seen me astride the little donkey. Here we wear the cassock, low shoes, long white stockings, baggy Arabic culottes and a shirt. *Voila, c'est moi!* I unbuttoned the cassock half way up, and clenched the flaps between my clerical legs and the donkey; so that I was fitted out with a divided skirt. The sight must have been funny! Today, an American asked a Frenchman if I was a Greek priest. How a shaggy whisker changes one! As I learned to ride in the years gone by, I took the lead with ease. I climbed the great pyramid of Cheops. Each step is about three feet up. I had to tuck up my skirts and pay no heed to my graceless togs. At the top, I sang the Star Spangled Banner, and took possession in the name of my country. The visit to the inner chambers was more difficult. We had to climb up steep and slippery passages, and to slide down several rock-beds. The engineer in charge of the excavations is an American convert and treated us most courteously.

Some day I will hold forth before the tribe, and will tell the wonders of Egypt.

His manuscript on the Jesuit Press in Beirut describes the school-work formerly conducted by the Society throughout Syria, and the splendid achievements of the Imprimerie Catholique in supplying books for these schools, and in printing the vast amount of scholarly works from the pens of such well-known Orientalists as Fathers Cheikho, Salhani, Rabbeth, and a host of others. The skilful work of the two lay-Brothers, Elias and Antun, in building up the reputation of the Jesuit Press of Beirut are also given high praise.

Writing of American Protestant forces in Syria, he explains how the popular attitude towards religion has in actual fact identified the official Consular forces of the United States with the work of Protestantism. In Syrian eyes, French officials have for ages been the official protectors of French religion and of French interests, i.e. Catholic religion and Catholic interests. In recent times they have in the same way come to identify American officials with the protection of American Protestant religion, because the chief American interests there are Protestant schools and colleges. Therefore, in the eyes of the natives, the American religion is the Protestant religion. It is unfortunate that no American Catholic missionaries have come to labor amongst them.

The zeal of his defense of Catholic truth, in later life, is evidenced even at this time in all these articles. Speaking of the difficulties encountered by the Jesuit University of Beirut, he says:

> The Syrian Protestant College had been established by the American Presbyterian mission, and was the chief instrument of Proselytism in Syria. The brave-hearted Father Monnot conceived the idea of fighting American Protestants by the aid of American Catholics. He therefore journeyed from Beirut to New York, stayed in the States two years, and collected the handsome total of $200,000. . . . The first and principal building of the Jesuit University was completed in 1875, and is even today the most substantial and artistic structure of its size in the city of Beirut. In the Seminary only forty or forty-five Seminarists are supported at a time. Ten times that number are presented by the Bishops of the different rites, but the exceedingly limited resources of the Seminary will not allow the support of more . . . If Catholics of the United States realized the wealth and influence of the American College of Beirut, i.e. the Syrian Protestant College, they would do something to help the famished Oriental Seminary of Beirut, every graduate of which is by his piety and learning and celibacy a strong checkline against the powers of Protestantism. . . To combat the same forces, the medical school was founded in 1876. By a strange disposition of Divine Providence, the man who most effectively favored the project and secured from the French Parliament the vote of 150,000 francs for the construction of the medical school was Gambetta, then Minister of Foreign Affairs, who himself added

25,000 francs for the equipment of the physical laboratory.

French assistance also made possible the establishment of free schools in all parts of Syria, conducted by the Congregation of native Sisters which the Jesuits have founded. The mission provides these poor Sisters twenty francs apiece per month; and with this paltry sum these heroines of the Lebanon live and thrive. True, the Superioress is cook and teacher at the same time, but she is none the poorer cook or teacher or Superioress for all that. At Ba'albek, for instance, the splendid American Protestant school building now has only six little girls, Moslems all; the rest of the little ones have gone to the Sisters' school . . . Such would be the history of every Protestant school in Syria, if there were money enough to carry on Catholic schools. But the closing of Jesuit schools and churches in France has led to the closing of two-thirds of the free schools of the Jesuits in Syria. It will take years of toil to regain the ground lost to the Church in Syria by the persecution of the Church in France.

CHAPTER X

IN PALESTINE AND SYRIA

A letter to his brother John, written on December 5, tells of his first plan to visit Jerusalem and the Holy Land about Christmas time, so as to have the opportunity of saying the Christmas Mass in Bethlehem. But this plan had to be changed on account of the prevailing bad weather, which would make the landing at Jaffa a hazardous venture, and in any case, would spoil the roads in Palestine for convenient travel. The following letter refers to this change of plan:

Université S. Joseph,
Beirut, Syria, 24 Jan. '07.

My dear John,

Many thanks for yours of the 30th ult. From this I see that you did not receive my letter of about the middle of November. It had a sad tone to it, and told of an attack of illness. The attack has long ago passed away. Since the beginning of December I have enjoyed splendid health and have been able to study very much to my heart's content. Superiors tell me that they often lose letters hereabouts. It may be just as well that you did not get my blue note; you would have worried without any cause.

I had to change my plan of visiting the Holy Land; the weather was very stormy and there was little likelihood that I could enter Jaffa. I shall go to Jerusalem during the summer, when the weather is more favorable.

Yes, my dear John, I have great confidence in never ceasing prayer. Often and often my breviary is for you and for Joe or for dear Mary or for Al or for Hugh. When I feel just a little bit the desolation of isolation, the prayer for one of you always gives me comfort and strength. I realize I am helping you, though two oceans lie between us. Internal experience makes me conscious, and external experience makes me aware that the character, no matter how strong, is not enough to save a soul. Today I have just read Matins of tomorrow's feast, the feast of the conversion of St. Paul. His war-cry was: "I, no, not I, but God's grace and I." I pray for God's grace for you.

I read the Storer papers with great interest [regarding

the Roosevelt and Vatican episode]. Don't be afraid to send me any paper you think will interest me. Just mark the article to save my time.

Lovingly,
Walter.

As for the visit to Palestine, an earlier opportunity presented itself in the following Spring, and it was planned to make the journey at the beginning of Holy Week of this year. Here he began a new diary, noting down, day after day, the interesting experiences that would be of value to him in his later work.

This trip to the Holy Land, or the pilgrimage as it may well be called from the spirit of piety and faith that pervade the record of his journey, was begun on March 25, 1907. With a party of students and priests from the University he boarded the steamship *Portugal* in the harbor of Beirut, at 4 p.m., and arrived at Jaffa at 7 o'clock the next morning. A single expressive statement marks the arrival: "By 8.30 Mass in the Holy Land!"

They had time to view the interesting sights of Jaffa, and in the afternoon began the journey to Jerusalem.

> We were on historic ground all the way. How my heart beat with joy as I entered the Gate of Jaffa and tread the streets Our Lord had tread! *Laetatus sum in his quae dicta sunt mihi; in domum Domini ibimus!* With what mingled joy and sorrow I entered today this Holy City. How good Our Lord has been to give me the chance to learn more and more of the life of our Life in *Quds*—Jerusalem, the Holy City.

The following letter to his brother, written soon after his return from this pilgrimage to the Holy Land, gives a graphic survey of the entire journey and furnishes interesting details of the Palestine of twenty-five years ago. From all recent accounts, however, conditions there are somewhat different now:

Université St. Joseph,
Beirut, Syria, 16 May, '07.

My dear John,

I came back a week ago from my trip of a month and a half in Palestine. Yours of April 7 and 16 awaited me. 'Tis too bad you subscribed to the *Sun* for me. I thank you very much for your thoughtfulness. Still I cannot get your money's worth of the investment. Joe subscribed to the N. Y. *World* (daily and Sunday), for me. I wrote him that I could not find time to make his subscription worth while. Fancy my bewilderment now with the *Worlds* and the *Suns* of a month and a half accumulated about me. Moreover, I expect to leave Beirut at the end of June for Germany. Father Provincial has not yet sent me definite word about the special university I shall attend. My letters will be forwarded to me, not my newspapers. Hence I hope you will switch the *Sun* off to yourself. You get great pleasure from the reading. Really I receive clippings and papers and magazines in superabundance now! Another reason why I prefer not to receive a daily (to say nothing of two dailies), is the temptation to waste time. I fancy 'tis a pity not to turn the pages over; and before I am aware, precious time has been lost to Arabic without any gain to make up for that lost.

Mr. Tierney, S.J., of Woodstock, sent me a Boston clipping with Hugh's photo in print. My itinerary, during the late trip, was made clear by postcards. The accounts of each day would require a letter. I was almost all day long, every day, either traveling or sight-seeing. The evenings were required for breviary and notes. These notes tell everything. Some day I'll give them to you. Had I written accounts of my journeyings to you, precious time would have been taken from sight-seeing. I hope to write up the entire trip in the near future for publication. There is so very much to say that I could not say it all in letters. My Arabic and Syriac would have to go to the wall entirely. Be patient, and in time you'll have all.

My first day in Jerusalem, Wednesday of Holy Week, was one of great devotion. I said Mass on Calvary for all family intentions. At every Mass, I go through the family, name by name, and tell the needs of each (even of little John and Charlotte) to Our Lord. This day your needs were more than usually insisted on. The Mass was beautiful, entirely in keeping with Calvary. My emotions leaped the barriers of will-control and it was with great difficulty that I read the Passion. Never, since my first Mass, was a Mass so much a joy to me.

The services of Holy Week I attended in the Basilica of the Saviour, which includes Calvary and the Holy Sepulchre. I had little of devotion. The sight of the lazy, gaping, gazing, talking, disrespectful Turkish soldiers, who were there to keep order; the lack of good taste shown by the tourists; the talking of Moslems and Schismatics—all this made me sad. I would not care to put on paper some of my experiences and impressions. I shall tell them to you later on.

Good Friday, during the last Stations of the Way of the Cross, I stood at the foot of Calvary, while the crowd was above, and listened to the closing sermon. Near me were some Americans who passed remarks on the bad taste and the cheap and tawdry decorations that were crowded together in the Basilica. I agreed with the Protestants, though *their* bad taste displeased me. Soon their talk went further, and poked away at Catholics. A lady (?) remarked: "Yes, they think it is all right to commit any sin they please, so long as they confess it." This was too much. I could hide my English tongue behind my Syrian whiskers no more. "I beg your pardon, sir," I said to the gentleman (?) of the party. "May I have your kind permission to say a few words in the cause of justice?" He was dazed for a moment, but found himself soon. "Well, what do you want?" "I wish to say that I have heard a remark which has cut me to the quick. It was just said that Catholics might sin as they pleased so long as they confessed. I assure you, sir, that remark is not true." He assured me that I had misunderstood, that he and his friends knew the contrary of Catholics, etc. The ladies craned their necks. I gave a parting shot. "There are certain doctrines a Catholic would die to save. Confession, sir, is one such doctrine." Off I went, happy to have done something for the dear Lord on Good Friday.

All that I saw in Jerusalem you could not ask me to write up. Every day was full of experiences. I said Mass for the family in all the great shrines—on Calvary, in the Holy Sepulchre, in the grotto of Bethlehem, in the grotto of Nazareth. I visited Ain Karim (site of the birth of John the Baptist), Emmaus, Reservoirs of Solomon and all the Bible sites near the Holy City. Jerusalem is the Holy City for what has been. It is now the dirty and accursed city. I would not advise Protestants or half-sceptical Catholics to visit the Holy Land. One's faith must be strong when one meets with conditions brought about by the caprice of the Turk and the *bachshish* of the Schismatics. The Turk has gradually sold Franciscan rights to Schismatics, so that now there is a great mix-up of rights and rites. Holy Thursday there was a row in the Basilica of the Saviour. Our service was delayed by a schismatic Copt who sang his Mass at top voice and was half an hour beyond the time allotted him. The Italian Consul intervened. I feared at one time the Turks would stop the Mass and put the priest out. Fortunately, only the congregation was ousted. This sight was part of my preparation for Holy Thursday Communion.

I was really pleased, when my trip took me out of Jerusalem. One excursion from Jerusalem was to Jericho and the Dead Sea. I had a swim in the sea. It is so salt as to irritate the pores of the skin. I was most uncomfortable for several days after my dip. On the following day, we set out for Jerusalem at 3 a.m. and reached it at 9 p.m. En route I shot at a jackal. It leaped. Maybe my imagination leaped. The morning was dark and we galloped along in our carriage. Our hurry was due to the wish of another Jesuit to start for Nablus.

Another excursion from Jerusalem was to Gaza, a most interesting tour of the land of the Philistines. The Fathers here think I was foolhardy to go among the Moslems alone. Still I knew no fear, and found they respected my Colt's revolver. This Colt's automatic had been brought to the University by some Christian Jesuit who served his time in the French army. At a town called Mejdel (near old Ascalon), I found not a Christian, nor a hotel, and slept in a vacant shop, a shed about ten by ten feet. Luckily there were a door and a key to the shed. During the night the Moslem youngsters threw rocks at my shed and tried to break in on me, shouted insults, etc. Still, by midnight all was quiet and had it not been for fleas and bedbugs and lice, I should have slept. In my trip I developed many a crop of human parasites. Am now clean. Our University is absolutely clean. I have not found a flea or any thing else here all this year. I speak not of those I imported from Palestine. I marvel, now that I have experienced the filth of the Orient. Why, even in hotels it is next to impossible to avoid vermin.

My stay at Mejdel was a memorable one. Two days later I stayed all night at Adjour. What a night! The only decent house was that of the sheikh. He refused to let me in. I lodged in a mud hut with a family of fellahin . . .

The most enjoyable days I spent in the lake country, near to the Lake of Genesareth, in Tiberias, Tell Hum (Capharnaum), Tabiga, Magdala, Bethsaida, Samakh. I tramped from end to end of the lake and enjoyed walking where our dear Lord walked and worked so much.

The home stretch was most interesting. After a trip

from Carmel to Mukhraqa (the site of the sacrifice of Elias), and thence to Caesarea, and back to Haiffa, I took the public carriage to Akka and followed the sea-board to Beirut. The towns of Akka (Acre), Sour (Tyre) and Saida (Sidon) are of interest chiefly from Crusader days, though the Phoenician fame has kept some traces in the old town.

Now I am most pleased to be back to Jesuit cleanliness, and far from the filth of the Orient. I say "far from" because the filthiness of Beirut does not enter into our up-to-date institution. Please keep the post cards. They will be of use to me, if not to you.

I have no word from Hugh or Al.

My health is splendid. Without fine health I never could have stood the hardships of my trip. Often I had only bread for food. Some of the bread I am sending Mother. Generally my food was that of the natives, crude, coarse stuff. The last day of the trip I walked from Saida to Beirut, forty-eight kilometers or about thirty miles. Quite an infantry march!

You and all the family and each of your intentions are in my every Mass nominatim. Pray for me. Miss M. I have prayed for. May our Lord care for dear Mary and help you all to bear and trust his will.

I am pleased that business has been good lately. May it be better! Leo Logan lately wrote me from Innsbruck.

Lovingly,
Walter.

The following letter refers again to the question of newspapers:

Beirut, May 18, '07.

Dear Joe,

I have yours of March 24 and April 6. Am right glad the Captain and you will be together. Be sure you steady him and he steady you. The *World* has come. But say, old fellow,—what about Arabic? Do you not realize you are putting me in the way of temptation to feed my imagination on the red-hot output of your fancy-free press agents? That red-hot output burns the brain cells, and unfits them for Arabic roots.

I was stranded at the outset, but am now fairly floated with news. This news comes in every form,—that of the frivolous *Messenger*, the side-splitting *Stylus*, the Gospel-truth New York *World*, etc. Where do I stand? Save me from conscience and don't subscribe any more. I hope soon to know my next year's destination. Shall write again. The Lord keep you good.

Lovingly,
Walter.

The following details from the diary will indicate somewhat of the interest and the variety of his experiences, devotional and otherwise, that were mentioned in the above letter:

March 27—Mass at Calvary, altar of the Madonna, about two meters from the hole in which rested the dear Cross. "Oh that my heart could utter," etc. The Mass was that of Wednesday before Holy Thursday. The readings from Isaias and the Passion were such a privilege to me. My emotions were out of will-control. The Mass was more difficult than my first Mass. Basilica, Sepulchre, Anointing Stone—sick from effect of emotions. Little of Calvary seen. Could not retain even fasting breakfast. Tenebrae. Two companies of Turkish soldiers to keep peace. One-half outside. Just inside door, a divan for Turkish officers, who smoke and chat there. One-half the company around the Sepulchre to keep off Moslems and those who were evidently not Catholics. The sight was very sad. The soldiers chatted and gazed about. Behind them was a pushing mass of every sort of people. There was no devotion in the beautiful ceremony for me. I left, went for third time to kiss the sacred spot of Calvary. There was a Brother, with Protestant women from the States. Thrust lighted tapers down into the hole of the dear Cross.

March 28—I visited Gethsemani, and said my beads for L. C. under an old olive tree. Returned with M. l'Abbe Fortescue and visited the Austrian Hospice. 'Tis a beautiful and palatial residence. After supper, at 9 p.m. to Gethsemani and Grotto of the Agony. The moon was full, the night was lovely. The pilgrims were pious. I said my beads for L. R. under the old olive tree, and said them once again in the Grotto of the Agony, for the family; and once again for the dear ones at home, when I made the sad pilgrimage up to the Holy City. Everything was in fine taste. There were no Moslems to make me unhappy.

The trip to the Dead Sea left him sick and in fever, and he had, therefore, to spend the whole of the following day confined to his room at the hospice. But the rest improved him, and next morning he was again able to resume his tour of the places of interest in the environs of Jerusalem. All are described carefully, and his observations, usually critical and terse, have to do with historical and archeological sites for the most part; they are here and there interspersed with his own personal impressions and outspoken opinions:

> April 5—This morning I said Mass in the Holy Sepulchre for the family. Before I got into the Sepulchre, two Russian women took possession. They did not know that I was pressed for time, and how their devotion interfered with mine.
>
> April 6—To Bethlehem. Said Mass at the Crib at 4 a.m. and heard Solemn High Mass at 7 a.m. My Mass was quiet and undisturbed. Even the Armenian chant did not cut me out from my devotion. I was very happy! Visited the Grotto of the Milk, walked past St. Joseph's house (?) through the Field of Booz and the Field of the Shepherds to the Grottos of the Shepherds—about two miles away. The Field of Booz is very fertile lowland, just the place for Ruth to toil. The little Ruths of Bethlehem are the sweetest little ones I have chatted with in the Orient. The town and neighborhood is thoroughly Christian. The Grottos of the Shepherds are caves such as are found everywhere hereabout. Probably tombs were here. There are bits of mosaic not yet unearthed. Maybe a crusader's church was here. The walk back to Bethlehem was very hot. From the road back to Jerusalem, one may see the cistern whereof David wished to drink.
>
> Bethlehem is a very Christian town. My joy today no words can tell. You are each of you in that joy. These family Masses I say for each of you by name; in each Mass, I tell the needs of each.

As was mentioned in the letter to John given above, he had occasion at this time to travel by himself, with but a single guide, to Gaza and the land of the Philistines. Though the journey was very useful and necessary for his purposes, his experiences on this trip were the most disagreeable of any that he had encountered in all his travels. The details that he gives in the diary are almost unbelievable. When one remembers the young dandy of a few years back, and how keenly sensitive Father Drum always remained to dirt and uncleanness of any kind, it is impressive to note the nonchalance with which he speaks of the uncouth discomforts of travel in the Orient. Nothing short of the highest spiritual motive could steel a man's soul to such privations, and it is a matter of admiration to know that his refinement and delicacy were never dulled by these experiences. His perceptions of beauty, and his reverence for the Holy Places were never blurred.

April 11—This morning we started for Gaza at 6.30 and reached the town about 10.30, after a painful walk. My horse's leg was very stiff [he had encountered a hailstorm the preceding day and his guide had neglected to attend to the animal]. On the way to Gaza, I passed a troup of camels that carried bits of beautiful white marble columns (from the ruins of Gaza) which had been wedged into quadrant pieces. The road near to Gaza is very sandy. The scenery is beautiful. The ground is very fertile. Huge fields were red with the poppy-like flower which the natives call *warda ahmar*, red-rose. This flower is common everywhere in Judea, never so common as in the plain of Sharon and in this land of the Philistines. I wonder if the Rose of Sharon were like this poppy!

In the twelfth century church of St. John Baptist, are some grey, veined marble columns, beautiful relics of antiquity. One wonders that these beautiful marbles and granites were brought such a distance when transportation was not very easy. Now the miserable hovels of modern Gaza tell their own story. While I looked about the town I thought of the glories of the past, the invasions of Egyptians, Assyrians, Jews, Syrians, Persians, Greeks, Arabs, Mamelukes, Turks! No marvel then, that Gaza's glory is gone!

This trip to Philistia proved very dangerous to his health, for on his return to Jerusalem he fell sick from the effect of poor food, the want of sleep, and the weariness of the journey.

After his recovery there were other expeditions in various directions in the neighborhood of the Holy City. He records his appreciation of the courteous hospitality of the Salesian Fathers everywhere. He visited many Biblical sites, and found especial enjoyment in his trip to Mount Carmel. From Carmel, the diary continues:

April 19—We made Jennin (on the mountain) about 11.30 a.m. As the plain of Esdraelon came in sight, it was lovely and green. At Jennin, I bought four loaves of bread, eight oranges, some *halawi* (sweets). This made dinner for us three. I could not get any cognac in Jennin. *Haram!* The Moslems were shocked when I said that cognac and even *Araq* were not *haram* to me.

I came into Nazareth at 5.30 p.m. and put up at Casa Nova. The English pilgrims (chiefly Irish) were there. A lady tried to talk French to me. I answered in English. She turned out to be a Miss Horgan of Macroon, a first cousin of Mother Dosithea, who is also Mother's cousin. Canon Keating of Birmingham was in the party and was very nice to me. I became very thick with Father Cooksey, an ex-Jesuit, whose brother is in Demarara and a Jesuit. The priests and people were most friendly. Many took me for a Syrian priest. One lady came up and told me she had heard me say Mass according to the Maronite rite, was so pleased, etc. She had heard another. Another lady had heard me start my Mass, but had left on finding it to be Latin, etc.

After dinner, visited the Mountain of Precipitation. On my way back I met a boy with some camels and took a ride for two *metalliks* (pennies). I dismounted without forcing the camel to his knees. The shaking up of the stomach makes camel riding disagreeable.

April 21—This morning I saw the antiquities of Nazareth. They are not many. In the afternoon, a Brother took me to the Mensa Christi. Later on I saw a wedding. I had met the groom's procession about noon, before the house of the bride. *Araq* flowed freely. There was much *fantasia,* as the Syrians call this nonsense, singing and dancing and general hullabalooing.

About three p.m. this procession reached the church of the Annunciation, and stayed outside to await the bride. In half an hour the bride came walking slowly, all in white and heavily veiled. The women who were with her, clapped hands and sang in very nice time. . . . The Franciscan read the service in Latin and the address in Arabic, and ended with the *Barake.* Everybody answered *Imbarake.* I understand the *fantasia* goes on a whole week before the wedding.

April 22—To Naim, Endor, Thabor. The tombs at the east of Naim show whither the procession was going, when Our Lord met the poor widow. I did not see the grotto of Endor, which is said to have been the cave of the pythoness.

April 23—Nazareth to Tiberias. Mejide. Saw Cana en route. Great pity the Franciscans built their new church at Cana without making proper excavations for ancient Basilica. The place of purification of the Jews was most interesting. The tradition seems good in favor of the spot on which rested the *hydriae* . . . . We saw the site of two multiplications of bread. I do not think the *Qurna Hattin* is the Mt. of Beatitude. 'Tis too far from the sea. I saw all of Tiberias today. Again and again I reached the sea, but failed to find a walk by the seashore in the village. The town is very Jewish in appearance.

April 24—Mass in the church of St. Peter, on the (probable) site of the *Amas me*, etc. 8 a.m. to Magdal, Tell Hum, and back to Tabiga. The ruins show that there was a very beautiful synagogue. The large limestone columns and bases are of pure Corinthian. The capitals are of Corinthian also, with acanthus leafing. Father Biever received me most cordially. Found Father Fonck and Doctor Rendiz at Tabiga. Father Biever is brother to a Jesuit in New Orleans.

April 25—To Safed with Fathers Fonck and Rendiz and Daoud. Father Biever gave us horses free. Dined with Austrian consul's family. Saw old fort at Safed. The Mariamettes brought out their children, who made speeches at us and gave us flowers. Compliments *m b'arif*, etc., etc. Yesterday, I ate fish cooked on open fire by Bedawin. One Bedawin said to me in Arabic verse: "Eat fish, drink water, when you're hungry, come to me!" The fish tasted very well. Very likely it was cooked just as Our Lord cooked the fish for Peter.

Several pages follow with description of various Oriental customs, one of which has to do with the part ownership of a pure-blooded Arabian horse. There are usually twenty shares to a horse and one may buy the use of the animal by buying five shares. One who has the use of a horse is said "to have the bridle in hand." The customs differ on both sides of the Jordan, and once Father Biever in crossing over had to give up his horse to the care of a partner.

April 30—At Carmel. Received well. Walked with Father Bernhard all about Carmel ruins. Too bad the Fathers lost much property to German Protestants twenty years ago. They bought property from the Sultan for a song. But they had not fenced in, nor built on, nor cultivated property. Process cost them 200,000 francs. They lost. Germans came in and bought land-quit claim 5,000 francs.

My Mass this morning was full of tender devotion. Our Lady was very good to me.

May 1—I am now writing at Aleka. In a church below my window, a priest is tearing his throat and bellowing like a mad bull. I wonder if Arabic calls for such fiendish sounds! . . . . Wrong again! No priest, but an Imam! No church, but a cafe! So the Imams howl to Moslems!

May 3—We reached Muhkraqa on Mt. Carmel about noon. Muhkraqa is an ideal spot for the sacrifice of Elias. The fire might have been seen from all the country round about. From the roof of the present chapel, one has a magnificent view, the best I have had in Palestine. The plain of Esdraelon was set before my eyes more extensively and clearly than from Thabor. Mts. Jennin and Gelboe nearby, and further on, the triple peaks of Little Hermon, the hills in which Nazareth rests, the portion of Esdraelon that runs towards Khaiffa, the rolling country between Muhkraqa and Caesarea, every hill for miles and miles could be studied.

The entry for this day contains a description of a Syrian meal—the Lebanon bread or *Khubz tannur*, "the finest bread I ate during my trip," the other dishes—pigeon, chicken, *zibde, jibn, kibbeh, baide.* "Big crash towels were served as napkins."

In the evening, I supped with a young man, a teacher from our college of Beirut. A low stool or table, that stood six or eight inches from the floor, held a plate for each and viands: cold broiled chicken and *zebrach*. This latter is a Turkish word for stuffed vine leaves. The stuffing varies, generally rice and chopped beef and seasoning (sometimes with mint leaves). A dainty *zebrach* is that stuffed with rice and chopped quail (*firre*). The viand is eaten leaves and all. It is very common among Maronites of Lebanon.

May 4—At Caesarea, after prayer and reason, I decided Our Lord would allow me to say Mass in the shop. I fixed up the counter. At best, conditions were wretched. I tried to make up in reverence for the lack of cleanliness of the shop. My Maronite stood guard outside.

I left Caesarea glad that there had been none of the trouble I was often told to look for. To be sure, the Circassians were not cordial, but they did me no more harm than to ignore me when I made a courteous remark to a group of them.

I followed the sea-shore to Tantura. I had no desire to see the modern Jewish town of Zumarin. I crossed Nahr ez-Zerka (blue river), called by Pliny, the Crocodile River. At Tantura, the ancient Dor, one sees very little of the Phoenician ruins that St. Jerome speaks of. Even the few stones left of El Borj will soon be marketed.

The Carmelites treated me with very great courtesy. It was with difficulty I got the brother to accept my share of cost of horses from Khaiffa to Muhkraqa and back. I said Mass of Our Lady of Mount Carmel once again, with devotion, and went down to Khaiffa. Here I found Fathers Fonck, Merkle, and Gillitzer, and Doctor Reuding. The latter had decided not to go with me via Akka. The two Bavarians were too sick to venture the trip. I readily got a place in the diligence and made Akka by 9 a.m. We followed the splendid bay. The horses trotted along in the water the most of the time. The wet sand offers good resistance and the interference of a little water does not matter.

The Nahr Na'man, River Bellus, was most interesting for the murex found plentifully there by the Phoenicians. In a cyst or sac in the throat of this little shell fish was found the Tyrian purple. Here too the Phoenician glass was first made.

May 5—I went to the Franciscan Church at Tyre, and there met the teacher of the Catholic school. He treated me very nicely, and showed me what Oriental hospitaliity is some times. After supper, coffee and chat. I never smoke. People wonder at this. I had to go over and visit the cousins of the Mu'allim. A great family awaited me. *Ahlan Wesahlan* (greetings) with a vengeance. Songs. I sang something serious. They were greatly amused with the joke of the thing. I sang something funny and accompanied the song with face action. The good people roared. One old woman became almost hysterical with laughter. And all the while I was trying not to scratch away the creepers and jumpers that swarmed in these huts. Soon May devotions began. The people were very devout. Most remained in a squatting position. The beads and litany of Our Lady and several hymns made up the prayers. Candles were lighted in honor of the Blessed Virgin and of *Mar Jorgios* (St. George), before their pictures. Incense burned before Our Lady's picture.

May 7—I reached Saida (Sidon) at about 1 p.m. The Jesuit Fathers insisted on my taking dinner. At about 3 p.m. I visited the town. What amazes me is the great number of priests and bishops in the Orient, and the need at the same time of Latin missionaries! Here there are a Maronite bishop and his priests for about a thousand faithful; a Melchite bishop and his priests for another thousand; for two hundred Latins there are the Franciscans and the Jesuits.

In the afternoon, I went out to the Asclepeion. On the way back a great storm began to gather. I rushed to keep ahead of it and said my Magnificat that Our Lady protect me. She knew that the people needed the rain, but I did not. Just as I finished my Magnificat, a Syrian passed me. *Massikom bil khair* (Greetings your anointing). He gazed at me and in Arabic asked if I were a Jesuit. Mysterious how he made me out in my American cassock. He turned out to be an old boy of our college at Beirut, and his son is now at the Jesuit college. He invited me to his villa near by. The rain was on me. In two minutes I should have been drenched. Instead I ate oranges at the country home of M. Feirant. He asked me if I could tell he was not a Frenchman. I was like little George Washington. We walked to Saida together. In the evening, discussion, did Christ visit Saida?

From Sidon, Father Drum returned to Beirut, arriving there on May 8, in good health and feeling well satisfied with the success of his long pilgrimage. He remained in Beirut till the end of classes, as the following letter indicates, in hourly expectation of word from his Provincial with regard to his work for the following year. Instead of merely marking time, however, the indefatigable traveler made plans for another tour of study and exploration. This was to be the most difficult and in some ways the most interesting of all his journeys:

Beirut, Syria, June 23, '07.

Dear John,

Yours of May 25 and June 5, with Jane's letter, are come. May God grant mercy to the soul of poor old M. . . . I had expected to leave for Germany at the end of this month, but have received no orders from Father Provincial. There is evidently some hitch. Our classes end this week. As I am to stay here longer, I shall take advantage of my stay for a trip to the Cedars, to Antioch, and maybe to Palmyra. My health continues to be remarkably fine. The heat is never unbearable in the shade. The thermometer registers about eighty-five all day long. In the sun, the heat is pretty bad. I shall be careful during the trip. There is not much danger as I shall be on horseback most of the time.

. . . May Our Lord keep you in good health and bring you safely over the present crisis. For three weeks I have missed two or three papers a week . . . I have lodged my protest. Nothing will come of my action. 'Tis the Orient!

Affectionately,
Walter.

But the itinerary that was finally decided on was quite different from that indicated in the letter just quoted. The diary, though not as copious in this journey as on former occasions, yet gives the names of the cities and towns that were passed through, and adds an interesting detail here and there. The trip began by water. Six priests made up the party that boarded the Italian liner *Singapore* from Beirut, on June 29, 1907, arriving the following morning at Lernaka, on the island of Cyprus. They were welcomed by the Franciscan Fathers and were shown over part of the island. The *Singapore* then made for Mersine on the Asiastic shore, which was reached the next day. A characteristic sentence appears in the record for this day: "July 1—Sea-sick as yesterday, but managed to say Mass each day." A carriage took them to Tarsus of Cilicia, scene of the birthplace and boyhood of St. Paul. The following excerpts are selected from the diary:

July 4—The town of Seleukiyyeh seems to have been largely carved out of the solid rock. Houses were built right into the rock. With wonder and pleasure I followed the tremendous tunnel. What energy was expended on this waterway of the Seleucides! Was it merely a waterway? Maybe it served some strategic purpose. At 4 p.m. I started alone for Antakiyeh. The night was dark, pitch-dark, after 8 p.m. There was no moon. The path was often invisible. I had to light matches and to follow my compass. At length I oriented by means of Venus, who stood just above Mons Silpius. She proved a good guide. My guardian angel brought me to Antioch safe at 10 p.m. Hotel keepers were astounded that I had come such a distance alone and in the dark. The natives have a child's fear of robbers.

July 5—At Antioch. Went out to Beit el Mai (home

of the waters) or Daphne, a beautiful spot. The dashing waters and rich vegetation charmed me. There is little left of the Temple of Apollo.

The entries for the next few days are unsatisfactory because he was not able to see the places and ruins that most interested him. The wishes of his companions had to be consulted, and they were for following the ordinary beaten paths of the guide books. His own taste was for the sights and ruins of classical and Christian antiquities that were not to be found in "Baedeker"—another instance of that greediness to make the very most of his opportunities, that was so characteristic of Father Drum. On July 9, they reached Haleb (Aleppo):

By far the most interesting thing in Haleb is the splendid collection of the Marcopoli family. Its china, Haleb pottery, mosaics from the Euphrates, Tudmor busts, and Roman-Greek statues, make a collection worthy of days and days of study. Mr. Marcopolis gave me an arrow from the Kala'at as a souvenir. In Haleb, the roofs and terraces are connected one with another,—at least in Haleb's oldest quarter. The Marcopolis may visit their relatives and friends of the neighborhood without leaving the roof. Roof gardens are splendidly arranged here. In the Marcopoli garden are orange trees that bear fruit. Haleb is the only town in which the roofs and terraces are such as the Arabian Nights and other such tales lead one to fancy. The Persian rug trade is still a Haleb trade. What gain between the merchants of Haleb and the U. S. millionaires! The real Persian rug will never change in color and will last hundreds of years.

July 10—Left Haleb about 5 a.m. Reached Homs 1.30 p.m. Started in at once to make ready for Tudmor (Palmyra) trip. French Consular agent was very courteous. He lives in the house of the famous Lady Digby, who married a Bedawi Sheikh, lived among her husband's people and aped their ways.

Visited our schools at Homs. Ours had forty schools here before persecution in France. Now have eleven schools. Homs is ancient Emessa.

On July 12, began the long difficult journey to Palmyra, out in the Syrian Desert. This part of the diary consists of mere fragmentary jottings, written down in the intervals of rest on the route, and indicating briefly the changing scenery of the country through which they passed, the distant view of the Lebanon range in clear, gray outline to the west, the nature of the soil, the springs of water at rare intervals, and the few uncertain ruins that were met with on the journey:

> July 13—The desert was parched and dreadfully hot. The soil is not sandy, but argillaceous.' A few shrubs grow here and there, such tufts as only camels eat. The soldiers know nothing of ruins we passed, save that robbers lurked there.
>
> July 14—Reached Ain el Fuaris (Spring of Horsemen). The water was bad, sulphureous. The fagged out horsemen reach this spring ready for any kind of water, and think little of the sulphur smell. A goatherd brought us sour milk, which was very tart and refreshing. This *leben* gave me a headache. I am here with full protection and letters of recommendation and what not. I was afraid of being put back as were Dominicans and others shortly before, without having seen anything of Palmyra. Ruins of Temple of the Sun, the Grand Colonnade, and many ancient tombs. The heat was next to unbearable. I climbed to the top of most of the towers. In the best preserved, the fluted columns are splendid.

The return journey to Homs lasted three days. He had been booked for Trablous (Tripoli), for July 19, but found himself too tired to undertake the journey, and therefore rested that day, and in the afternoon started southward for Baalbek instead, where he was given lodging in the convent of Les Dames de Nazareth. "Thirty are coming tomorrow," they said, "why not one tonight?"

> July 20—Poor dear Sisters were up to hear my Mass. I left Baalbek at 5.45 a.m. I had a splendid, strong horse for the climb to the Lebanon range. Reached Deir el-Ahmar in two and a half hours, Aineita in two hours and three-quarters more, and rested about an hour at Ain Nebi'a, a glorious spring. The water was so cold that even then, in July, it was painful to keep the hand therein. Great volume of flow.
>
> The climb up Jebel el-Arz, Mount of the Cedars, was circuitous and tedious. I had to cross snow. My horse ate of the snow with zest. I walked down the mountain to give the horse a rest.
>
> As we started the descent, what a glorious view! Before us, Becharré perched above the beautiful Wadi el Qadis! To the right, the Cedars! There they were in all their glory and age. The mountain itself is barren,

and in the midst of all this barrenness arose the Cedars of Lebanon.

I stayed three hours at the Cedars. Met a young priest of the African missions, who showed me everything. On one very old tree are the names of Lamertine and his daughter Julia. In another tree is the cell of a monk of old. Nearby is what the Maronites deem a miraculous fountain of water. The story goes that this monk was unwell and could not get down from his lofty perch. He prayed and the Lord sent him water. In a little basin in a crutch of the tree one always finds water.

I went to the Carmelite Convent of Mar Sarkis (St. Sergius). The Fathers were most kind, despite their great poverty. There is water everywhere. Some of the water is calcareous, but most of it is delicious. I said Mass at 4 a.m. and took a cup of coffee. No bread in the Convent! Started the descent to Becharré, ascended mountain, past the Cedars, dismounted and walked up Jebel el-Arz. View of Wadi el Qadis very fine. There was not a cloud in the clear blue sky. The air was so pure and clear that everything seemed near at hand. Made Baalbek by 3.15 p.m. and took train back to Beirut at 5.30 p.m.

The following letter was written to John a few days after his return to Beirut:

Beirut, 30 July, '07.

My dear John,—

I have your letters of June 4, 17, and 30. My delay to write has been due to a trip up north and to odds and ends that awaited my return. There were no postals for sale except at Haleb. I sent you postals from Haleb (Aleppo) and from Antioch.

On my return, I found orders to go to Innsbruck, Austria, to study under care of a Jesuit there. I am now not in the hurry that was mine when I expected to atttend lectures in German at some German university. As I wish to rest after my travels, I shall not leave here till Aug. 24.

My health has been remarkably good all this year, maybe on account of my travels. I study very hard when in Beirut, but am out of Beirut as much as in it.

The heat is most unbearable outdoors, but bearable enough indoors. The temperature in the shade is wonderfully fixed. Morning, noon, and night, the temperature stands at eighty-five. The lack of fresh evenings is to be regretted; still, I enjoy life here and can sleep. Our building is remarkably clean and free from the pests of the Orient.

I regret that your health is not good and hope you will improve. The war talk has set me praying the dear Lord to ward it off. You are necessary for Mary's care. We cannot let you go with the Ninth Massachusett. Hugh is most successful. He deserves the success, for he has sterling qualities. I am praying for all your intentions.

The bread for mother was a sample of the sort I live on in my trips. An outline of my last trip will be of interest to the family.

There is a short summary from the diary itself. The letter continues:

After a few days of rest I made a trip of two days in the Lebanon. Shall make one more such trip and then leave.

I am now rested and pretty well cured of flea bites. I still use vaseline on my poor, bruised legs.

May Our Lord bless and care for you all.

Love to all. Ask Joe and Hugh to write to me.

Lovingly,
Walter.

The Fathers and Scholastics of the community of St. Joseph's were nearly all in the villa of Bikfaia, or in the College at Ghazir. "In either place, one had the cool mountain nights and an occasional breeze by day. I spent only a day at each town, as I had seen the Lebanon well enough and wished to do some special work before leaving Beirut." Word from the Provincial had arrived while he was absent on the trip to Palmyra. The following letter to one of his cousins, a Sister of Notre Dame de Namur, speaks of his destination and coming studies:

Beirut, Syria, 15 Aug. '07.

Dear Sister,

When I wrote you last summer, I had no intention to cross the Atlantic and the Mediterranean; such intention came later and was due to the intention of the Provincial.

My year in the Orient has been very fruitful. I was about a month in Egypt, a month and a half in Palestine, and a month in Northern Syria, so as to learn the land and its peoples and their ways. These travels were sandwiched in between the hard task of Arabic and Syriac and Hebrew roots. I hope you will digest the mixed metaphors of a hurry-note.

My health has been excellent because of the trips. I have been able to grind away at high pressure during the intervals between explorations.

The holy sanctuaries were, of course, an occasion of much joy. I shall never forget my first Mass on Calvary. It was the Wednesday before Holy Thursday. I had reached Jerusalem the night before. The Mass has ever meant much to me; for, after that day, there is no Mass for me during three sad, Massless days; and again, the Mass of Wednesday of Holy Week contains the choicest predictions of Our Lord's sufferings, the great chapter of Isaiah, the Passion according to St. Luke. But I must not be wordy. That Mass was all the world to me! It took me long to say it, but there was no press. I never had felt, nor ever shall feel as I then felt the comfort of the Passion of Christ!

There is no use of my trying to tell you of my travels and studies and experiences; they would take more time than I can give and will all keep fresh in memory till I return to the States.

I leave 24 Aug., for Universitätsstrassen, Innsbruck, Austria; to go on with my Biblical and Oriental work.

If you again meet Reverend Brother Peter, pleave give him my very best wishes; I have a fond memory of his help and kindness during that Retreat which I gave the young Brothers [at Pocantico Hills].

I thank you for your kind words anent my writings. Oh, no! I shall write again for the reviews; but my field of work will be Biblical and Oriental. All my fond hopes to write apologetics are nipped in the early bud.

I enjoy very much the Bible studies I am now at.

Please offer my hearty sympathy and promise of prayers to Sister . . . .

I have no retreats this summer. I miss very much the chance to use my priestly powers. But God wishes me to sacrifice even the desire to aid souls by direct intercourse. I am glad to make the sacrifice, as there is very little that I may offer the dear Lord. Pray for me.

Devotedly in Domino,

Walter S. Drum, S.J.

CHAPTER XI

GREECE AND ITALY

True to his usual custom, Father Drum began immediately a new diary of his travels in Greece and Italy. It was not necessary to reach Innsbruck before the middle of October, and he therefore decided to use the intervening time in a long leisurely tour of study and exploration. The first part of the journey, to Constantinople and Greece, was cared for out of the munificent funds provided by the French Government for the traveling expenses of students and professors of the University of St. Joseph. The trip itself is interesting for the many experiences that Father Drum was careful to note as he went along, and we can see that on this as on other journeys it was the scholar and the artist, as well as the pilgrim-priest, who reveled in the wealth of Old World glories and garnered richly of the inspiration that may always be harvested from the treasures of Greece and Italy. The devotional, the artistic, the classical, had ever an appeal to his interest, while the lure of modern art or the curiosities that draw the ordinary tourist attracted him not at all. It is only necessary to select a few of his notes and jottings to see how thoroughly he enjoyed and profited by this tour of the classical lands of the West.

25 Aug. 1907 Sunday. On the deck of the *Saghalien.*

We are moving along evenly enough at about fifteen knots. Cyprus is in the distance. We passed along its southern coast last night.

I said Mass at 6 a.m. My companion said his Mass at 8.30. The Captain was most courteous and gave me permission to use the fumoir at any hour I wished. Very few were present at Mass. Only men—the women were most of them sick. I do not yet know the passengers. I am in soutane and whiskers, and they keep me at a distance. As I travel to Constantinople with a French Jesuit, delicacy demands that I do not Americanize myself all at once.

27 Aug. Tuesday, Smyrna. I was up at all hours, but slept on deck better than I could have in the cabin. As soon as the hour seemed not too unseemly, I awakened our *garçon* for the key of the smoking room, and said Mass

at 3.30 a.m., before the noise began. The smoking room is placed near a small freighter (a machine for hoisting and shifting freight); and Mass is very far from devotional when that machine is going. We came into port at 5 a.m. I spent about five hours on shore. I climbed the hill back of the city. The fortifications interested me much —they are not artistic, they aim at bulk. What interested me more was that I got along with my classical Greek. 'Tis too bad I am ignorant of modern Greek. By learning the pronunciation of the Greeks of today, I have fortunately been able to get along with classical Greek and Arabic. On my way through the Moslem quarter, a little negro boy followed me about for fifteen minutes, cried *pappas* (priest) and flung stones. My guardian angel kept me from the stones.

28 Aug. '07. Mass bright and early. Many Christian brothers received Communion. During the night, a Capuchin bishop died in the first-class part of the ship. He was very far advanced in years and was deranged by the motion of the boat. His remains were taken ashore at Dardanelles which we reached about 5 a.m. Constantinople at 4 p.m. I enjoyed the approach very much. An engineer from Paris pointed out the things of interest. The site of the city is unrivaled for beauty.

Here again as was the case at the first coming to Constantinople, there was great difficulty with regard to his passport, and it was not till late the following morning that he was allowed to land.

The boat was booked to leave the city at 3 p.m. I should not be able to see anything of the city. I decided to await the next steamer of the Messageries. Also wished to make our Consul get an apology for the shabby treatment with regard to my passport. After dinner, took a drive to Scutari to see the howling dervishes. The scene is diabolical and revolting. I left in disgust and shame,— so I missed the sight of the finale, when the imam walks upon little children and hell is set free from the depths of lusty lungs.

30 Aug. '07 Friday. I visited Pera, a cosmopolitan city of Greeks, Armenians, French, etc. I saw three or four old churches. That of the ancient Society was interesting. It was the first church allowed by the Moslems. Louis XIV wrenched the permission from the Turks.

31 Aug. '07 Saturday. At 9 a.m. I reached the Museum. I had seen the wonderful Sidonian tombs and was

most anxious to study the sarcophagi that Hamid Bey had found at Saida. The sarcophagus about which I spent most of my time is called the Weepers. Upon its four faces are the high reliefs of eighteen mourning women; no two have the same attitude; no two express the same grief; the eyes, the lips, the arms, the expression of the face, everything changes in every carved figure. The richness and variety and delicacy of the artistic work of these figures gave me as much pleasure as a piece of fine Greek poetry used to give me in the days that are no more. Another fine sarcophagus, called the Alexander, is more esteemed; but it has not the individuality and the intense intelligence that I admired in the Weepers. I shall have to use this in a sermon. It made me think. Why not bring it to make others think?

At noon I met the Consul General. He kept me an hour, treated me with the utmost courtesy, and promised to take action. He will demand apology and one hundred francs. He expects to get not a cent, but wishes to make a fuss so as to prevent future trouble. That is my only reason for insisting on an apology. I wish to pave the way for future priests.

2 Sept. At 1.30 I went to St. Sophia, and spent two hours enjoying the beauty of this grand old church. Unfortunately almost all the mosaics are covered over by cheap work, but the general outline of the church gives one much pleasure.

A few brief entries for the next few days chronicle the resumption of the voyage back to Smyrna, and then to the Peiraeus, which was reached on September 6:

Athens is a very beautiful city, a sort of vest-pocket edition of Paris; the wide streets and boulevards are a great pleasure to one who comes from the filthy and narrow ways of Constantinople. There seems to be considerable wealth in modern Athens.

Sept. 7, '07. In the afternoon, I betook me to the treasures of the Acropolis and its neighborhood. I went over the old ground with new pleasure. These grand old monuments never weary me. The Temple of Zeus, the Parthenon, I find new pleasure in them every time I see them, whether from near or from far.

Sept. 8, '07. This morning I devoted to the Mykenaian curiosities of the Museum. I knew the culture of the Mykenaian age by my study of Homer. Now to realize that culture in the exquisitely wrought cameos was like reading Homer for the first time. Some of the goblets of hammered gold are marvelously wrought and the inlaying of silver and gold shows a civilization that was very high and refined.

Sept. 9, Sunrise on the Acropolis of Corinth, Akro-Corinth! The view was glorious. The mist was lifted little by little, till at last there appeared above, the clear blue of the Ionian sky, and there lay, far below me, the gorgeous setting of the Morea, the Saronic Gulf, the Gulf of Corinth, Attica and Argolis. I was satisfied with my climb. I love the beauty of nature as seen from the heights of God's mountains. On the way down, I drank deep of the Peirean spring, the spring that gushed forth at a stroke of the hoof of Pegasus.

Sept. 10, '07. Mass in the hotel. How fortunate I have the portable altar. I would never have made these Grecian excursions without my Mass. One Mass is more to me than all Greece. But all Greece is much to me and I am ever glad to see part of it.

The Museum at Olympia. The restored groups from either pediment of the Temple of Zeus are wonderful examples of the best period of Greek art. But, oh, the Hermes of Praxiteles! What a glory it is. I had read ecstatic descriptions of the statue and almost feared they were Hellenistic ravings. The Hermes is like Niagara; it defies description—it is the one work of man that I have enjoyed most. Of course I speak of purely natural work. Every line is clean and seemingly perfect, and tells its tale of muscular and manly beauty, a beauty one sees no more in Greece.

Sept. 11, '07. At Nauplia, my supper was two mutton chops (price 30 *lepta*) a piece of bread and a glass of wine (5 *lepta*); in all, 7 cents. I shall not break the bank of the Province of Maryland-New York.

Monday, 16 Sept., 1907. The Captain refused me the use of the fumoir for Mass, but kindly offered me a cabin all to myself for Mass alone. I could not blame the poor fellow for his refusal of the fumoir. We had on board the daughter of the hero Clemenceau! I am with my former Rector of Beirut, and a fellow student, with four or five priests, a half-dozen Christian Brothers and two Religious.

Sept. 17. Naples. This and the following day were spent in explorations. On the 19 September, the miracle of San Gennaro, St. Januarius! It was a wonderful experience. We arrived at the church at 7 a.m. with a note to Mgr. Lezzi. At 9 a.m. we were admitted to the sanctuary. Outside in the church, the crowd was already packed like sardines, as well as in the Cathedral, to which the church of San Gennaro is joined.

Mgr. Lezzi proved to be our old friend of Las Vegas. He was mighty good to me; but told me that my own efforts would have to find me a place among the three hundred or more who were crowded into the sanctuary. "Never fear," said I. In fact, I was in the very front row and on top of the predella, and near to the Monsignore who held the reliquary. There was no grace, no tact, no delicacy about the arrangements. One had to push for ground, and hold the ground gained. But the occasion was one of a life-time. The sight of that great, swaying and shouting multitude was indeed inspiring. Their intense earnestness and absolute certitude that the miracle would be wrought appealed to me as a mighty instance of Divine faith. 'Twas wonderful and awful. Cold-blooded Americans might be shocked at their way of showing reverence and faith. By 9.30 the cries of impatience increased. "We will it! Blessed Saint, work the miracle! We must have the miracle!" I had no time for moral reflections nor for thoughts about good manners. They did not doubt. Their simple faith was never shaken. For sixteen hundred years and more, San Gennaro had caused his blood to liquefy on this very day each year. I knew he would not be wanting today. I saw the hardened blood that scarcely filled the phial. I watched closely for an hour. At last, at 10 a.m. the great moment came. The Monsignore smiled, and nodded to those near him; his assistants tossed handkerchiefs in the air; a shout of joy arose from the crowded thousands; the cannons boomed above the chapel; and bells rang out the joyful tidings throughout all the city. A great Te Deum was joined in by all in the chapel of the Saint. The miracle was wrought right before mine eyes! It was one of the great moments of my little life. My eyes were too dim to see further and my one thought was this: *Mirabilis Deus in sanctis suis* (Wonderful is God in His saints). My faith was strengthened. My heart beat with joy. I prayed the saint for you all, and especially for N . . . who has lost the faith. After I had well seen the phial filled with liquid blood, I left the great scene for Pompeii. On the 21st, my birthday, I said Mass in the church of San Gennaro, for all the family, and at 9.15 again saw the miracle of the liquefying of the blood. This goes on all through the octave of the feast. Today the liquid blood only half filled the phial.

> 22 Sept. Sunday. At 1.45 p.m. started for Rome. My heart leaped for joy as the train sped along the Alban Hills, passed close to the Appian aqueduct and rushed into Christian Rome. I am at the Gregorian University.

There were two weeks of tramping and sightseeing in the Holy City, the daily program being so carefully mapped out that he was able to see and study many of the wonders of Rome. But though he enjoyed, as only an art-lover and a religious student can enjoy, the treasures of Christian and classical antiquity heaped up in the city of the Seven Hills, with its five hundred and more churches and countless relics of bygone splendor, yet his comments show an independent judgment, and are not at all the conventional and hackneyed expressions one may find in any guide book of the Holy City.

> The church of St. Ignatius is beautiful, if one can only love the baroque. The altar of St. Aloysius is almost as chaste as I would have it. Today the Pantheon pleased me! there is not yet enough of barocco patched upon its venerable old walls to ruin the grand old lines. What a pity that I love only the simple Greek lines! It almost worries me to see fantastic and meaningless varieties of marbles and precious stones; they are almost as bad as brass or paper flowers set upon a delicate and snow white marble altar. St. Peter's is not severe enough for its greatness in size. Even now the mosaic restorations and additions seem to me to render the tout-ensemble painfully gaudy. Our much admired churches of St. Ignatius and the Gesu have left me in chagrin at their giddy gaudiness. What a confirmed old crank I am!

But the diary gives evidence of abundance of spiritual enjoyment and delight in these few days of his stay in the Holy City. There was the Mass each day at some one or other of the shrines of the Society, and the extremely cordial and pleasant companionship of the Fathers of the Curia with whom he sojourned. "Father General had not yet returned to Rome. The Fathers Assistant were simplicity and goodnaturedness personified. I was delighted with the community recreation, and never for a moment had a thought that I was among the big men of the Society." Such was the naive comment on his welcome at the headquarters of his Order. But a still grander favor than any he had yet received was in store for him before his departure from Rome. It has already been related at the beginning of this sketch, but the following details may be added:

27 Sept. In the evening I called on Mgr. Bisletti, the Major-domo of the Vatican, and obtained permission for a private audience with the Holy Father. The Monsignor chatted most graciously concerning my studies.

30 Sept. Monday. Mass in the room in which St. John Berchmans died. What a wonderful chance I have to satisfy my devotion to and love for our Saints! Last night a special messenger came from the Vatican to give me the enclosed document. [It was the printed form-letter stating the date and time of the audience.]

After an hour spent in the Sistine Chapel and prayer at the tomb of St. Peter, I presented myself for my audience with the Holy Father. At 11.30 a.m. the great moment came. I had passed all the guards and flunkies.

The Monsignori were most cordial. A Jesuit is much esteemed, I find, at the Vatican. Father Carroll of California Province was with me. We went into the visible presence of the visible head of the Church. I shall never forget that visit. It was another great experience in my little life. I kissed his hand and made to kiss his feet; he kept me by the hand and made me sit down. Then came a most charming chat of nearly half an hour. My studies were the subject. The charm and simplicity of that grand old man I shall never forget. When I could not bring to mind the Italian word, I gave it in French or Spanish or Latin; and the Holy Father put me the word I sought. I talked about you all and received a special blessing for each of you. I had prepared a picture to be signed by His Holiness. He wrote upon it a special blessing for mother: *Dilectae filiae benedictionem apostolicam ex corde impertimus...* (To my beloved daughter I grant my hearty and apostolic blessing.) He told me to thank mother for giving me to the Church and spoke very kind words for her. The Holy Father granted me some very special personal favors. Indeed the dear old man was so thoroughly simple and kind, that he touched my heart in a way never to be forgot. My love for him and enthusiasm are such as I can never describe.

Oct. 8. Tuesday. Mass in the Gesu at the altar of St. Ignatius. It was an affecting Mass of farewell to the father of my soul.

Mgr. Marchetti, of the Secretariat of State, a great friend of mine, is come back from his vacation. He called on me this morning. He was once Auditor [legal advisor] of Mgr. Falconio at Washington.

At 9.55 I set out for my roundabout trip by train to Innsbruck. 2 p.m. arrived in Assisi. Pilgrimage to the shrine of the great Saint was a help to me. In evening, started back to the station three miles away. No moon, and the stars were feeble and my compass served me not. I lost my way; my guardian angel found it. Plenty of time and to spare. In the waiting room of the station I gave a kindly sermon to some rabid anti-clericals. They took my words better than I had hoped for. I became a great friend of one who had been most violent against priests. My telling argument is ever an appeal to the liberty and esteem that priests enjoy in the land that all Italians look on as the ne plus ultra, the *Stati Uniti.*

Oct. 9. Wednesday. Loreto. I said Mass within the Santa Casa, the Holy House of Nazareth, which was brought hither by Angels. This little house is the object of fondest Jesuit memories. Sts. Ignatius, Francis Borgia, Aloysius, and John Berchmans and many other Jesuits of renown made their pilgrimage to the dear Lady of Loreto. My Mass was one of devotion and you were all in it and in my fervent prayers at this dear shrine. You see, this part of my journey is more a devotional than an educational trip.

Oct. 10. Thursday. Florence. All day today I have been visiting the beautiful churches of lovely Florence. They are in general far more pleasing to me than are the churches of Rome. The sixteenth century bad taste was ruinous to much that had been grand and beautiful in Rome. The abiding influence of Michelangelo, the splendid taste of the Medici, the wonderful painters Giotto, Fra Angelico, Fra Lippo Lippi, followed by Verocchio and his great pupils da Vinci, di Credi, Perugino, Botticelli, Ghirlandajo, and Fra Bartolomeo—all these influences and forces made for a standard in Florentine art that has given me more joy than I can ever tell. Il Duomo, Santa Croce, and Or San Michaele were my chief churches, tho' I saw all others that were of interest, the Annunziata, Santa Maria degli Innocenti, La Badia, Santa Maria Novella, Santa Trinità. They each have something of interest. All day long I've been running about full of enthusiasm.

15 Oct. Tuesday. Venice. San Marco kept me busy all afternoon. The mosaics of the twelfth and thirteenth centuries are so beautiful they made me wonder

how it is that Protestants so often refuse to admit the existence of art and science during the "dark ages." San Marco is rich, wonderfully rich, without being Barocco. Santa Sophia was grand and majestic; San Marco is grand and magnificent. These mosaics have brought me back to that love for Byzantine which Roman churches had caused to slip away.

17 Oct. Thursday. I leave Venice today. My memory of the city will be ever most pleasurable. A spirit of quiet and refinement reigns in this beautiful place. There is no clatter, no street noises. All trams and wagons are replaced by the light *gondola* and *barca*. The *gondolieros* I have found gentle, and not noisy like the hack drivers of Naples. One thing most pleasing is the piety of Venetian art. There is nothing pornographic in the palace of the Doges. The Venetian painters drew their high inspiration from the treasures of Christian history and thought. The Crucifix, the Christ in glory, the Madonna, St. Sebastian, St. Christopher,—such gave inspiration to Venetian art in the time of the Doges.

19 Oct. I was wide awake as we reached the Alpine divide, and the soft light of a beautiful moon made the Brenner Alps to stand out in clear and charming outline. We came into Innsbruck at 3 a.m. I waited in the station two hours and then came to my destination. My long and interesting journey is over. By the next day I am settled down to quiet work and am resigned to my whiskerless chin. My first task is to acquire the habit of speech once again. My start at Beirut was rather disconsolate; my start here is more so. Then I knew some French; now I know no German to speak of.

CHAPTER XII

INNSBRUCK

But it was not long before he settled down to routine work and took up his old studies with renewed zest. To quote from a long letter to a relative in Fall River, Mass.:

My new home "likes me well." I have not yet the knack of talking; it beats the Dutch! My work this year will be Old Testament exegesis,—i.e. interpretation. I shall take up Assyrian—my thirteenth tongue! For good luck's sake I must look out for another. The town of Innsbruck is beautifully set in the Brenner Alps. Round about us on every side tower the snow-capped giants. I am overjoyed with the beauty of the scenery. Lately I made my retreat at our summer-house, the villa, and was often distracted by the beautiful colors of these mountain tops at the rising and setting of the sun.

The following letters from Innsbruck give some account of his occupations there and the principle happenings of the year:

Innsbruck, Austria,
11 Nov. '07.

My dear Hugh,

Accept my most hearty congratulations on your appointment to the post of regimental adjutant. What a continual run of honors you have! Battalion Commissary and Adjutant, Regimental Q.M. and Adjutant! Why, you pick up everything in your way! I am very happy, dear Hugh! You deserve it all. Your superiors know that whatsoever you do, you do well and very well! You deserve great credit for all you have accomplished, especially as you were handicapped by youth and the short cut you had to take through your studies. You have overcome your many difficulties nobly. I am mighty proud of you, Hugh. I wish I could give you a good, hearty handshake. Mr. Ahern, Leo Logan, Fr. Tom McLaughlin (of your class at S.F.X.) are here and send congratulations. I hope and hope that you will next time reach the General Staff.

I am very well out here. My stay in the Orient has left me tender of skin, and the cold of this Alpine region goes through me with a piercing that you experienced on your return to the States. Still I'm careful. My studies are along the old lines, Biblical and Semitic. I've begun my thirteenth language—Assyrian, and a tough task it gives me. For good luck's sake I'll call Modern Arabic and Ancient Arabic two languages, so as to make Assyrian my fourteenth. There is no alphabet in this language. I am learning three hundred wedge-forms and angle-forms to start with. Once the start is made, the finish will be easy. The chief difficulty is to decipher the script. Here is a sample [a drawing of some of the letters follows]. Every combination of wedges and angles has a different syllable-value, and a special idea-value. All this work is hard for one of thirty-seven, but is in a grand cause. I must know Assyrian, so as to be an authority in my writings on the Hebrew text of the Old Testament. I do not suppose Mary or you envy me my job of cuneiform (wedge-form) studies.

Pray for me that my health keep good, and that I be a priest after the wish of our dear Lord; that my life be through Him and with Him and in Him now and for the rest of my days. Never a Mass goes by without a very special intention for Mary and Hugh. Your intentions are in my Mass by word.

I have written fully about my travels. You have doubtless seen my letters. Time does not allow that I repeat myself, and so please send this letter to Joe and John.

Thank you very much for your birthday letter. You were specially in my prayers and Mass the 19th Sept., when I saw the miracle of St. Gennaro (in Naples). Please write me M.'s birthday.

Love to Mary and you. Good-bye. Walter.

I talk German here. The recreations in German are not yet easy-going.

Innsbruck, Austria,
15 Nov., '07.

Dear John,

I have received yours of the 29th Oct., the *McClure,* Dooley clippings, and the Terre Haute paper. The Dooley sayings are refreshing. Let me have some more. They make me laugh. A laugh does me good, and helps me to study. My life here is so utterly monastic and retiring that I have very little to take my thoughts off Assyrian cuneiform ideographs and other Semitica.

Indeed, the Lord has given me remarkable opportunities to see the world and to broaden out with it. Traveling is most enticing. Still, my life work is to be such as not to give me very much more of the world-trotting, once I am settled down.

My health is very good. I have not weighed, but fancy I am picking up a few pounds. Studies are most enjoyable. Now and then I feel it that I am cut off from priestly work. The power to carve character, to knock the devil out, to help grace in—these wondrous priestly powers are pent up and I would like to let them out. But in after years, I shall have many a chance for confessions and contact with souls. The hard work and lonely work I do now will reap its fruit and harvest in a lifetime of usefulness, God willing.

Be sure to pray for me every day. Ask the good Lord to make me true to the graces he generously gives me.

Your intentions are ever in my prayers and Masses.

Lovingly,

Walter.

Innsbruck, 11 Dec., '07

My dear John,—

. . . I wish you a very Merry Christmas and Happy New Year. At present I am in very good trim. For two or three weeks I had a cough that worried me. I could not rid myself of it by medicine. At last I took to my old Kneipp methods,—had no fire in my room, filled my room often with fresh and cold air, breathed deep and full draughts of air, walked and used cold-water baths, etc. Results—I am in very good trim. I came from Syria weakened by the constant heats. My tender skin was in need of vigorous hardening; it got the dose and now acts well.

Thanksgiving Day the forty Americans (of whom seven are Jesuits) had a glorious banquet. I whooped it up with a stump speech! Heavens—what fun it was to talk! I feel like a chained hound! This was my first chance to talk since my sermon on the Atlantic; and was my first chance to talk to men outside of a church. What a difference! How easy to score points, to gain applause, to bring out a laugh, to be in touch with the audience! I wish we could have applause in the church. The encouragement given by the signs of approval is wonderful. We sent a telegram to the Pope, thanked him for the condemnation of Modernism, and received a splendid reply.

I pray the Lord to better your business opportunities. Please enclose these letters in the next letters that you send to the boys. In yours of the 24th you tell me that you declined to run for the school committee. I think it wiser that you keep out of that work.

My studies are as interesting as ever. German is not now the bugbear it was at the start. With love and best wishes, and the assurance of Christmas's and New Year's memento,

Affectionately,

Walter.

The Thanksgiving celebration mentioned in the above letter is still recalled as one of the happiest memories of American Innsbruckers,—there is a society of former students of Innsbruck, under that name,—and the event was chronicled in the Innsbruck *Korrespondenz Blatt*, as also in this country by the Louisville *Record*, which printed an account of the festivities and quoted Father Drum's speech on the occasion.

To a cousin, a school teacher in Fall River he wrote from Innsbruck as follows:

Innsbruck, Jan. 12, '08.

Dear ———,

. . . Don't mind my mixing up fun and earnest. I have no time to separate them by the proper transitions that you learned so well in high-school. The earnest is my sincere thanks for your New Year's Communion. I am so pleased. I want more. Now see what you have brought upon your good little self. Here is a great favor I beg.

On 2d Febr., I take my last vows. Word has just come to me. Now I wish you and the others to make a novena of *daily Communions* for me. I had intended that the novena end on my great vow-day. Now I am too late. Never mind! Start it at once! Your novena will be going on, while I go off. 'Tis a great event, Joe. I've been looking forward to it almost since you were born! You see, I wish to draw down God's abundant graces, to receive them in most powerful inflow. The more prayers and Communions are offered for me, the more will the dear Lord be generous in helping me. I do not wish to put you all to over much bother; and know that you will do what you can for me. But you, I have a special claim on; you I cannot excuse. If you think me inexorable and hard, then blame all the nice letters you have written to me; and the legislative and judicial and executive authority that you allow me to wield in your life of fishing and 'phoning.

I go into retreat the evening of the 24th.

Enclosed is the taper,—all that is left of it,—that lighted me through the catacombs of St. Calixtus. Maybe you are bleak and black sometimes. Then light the taper, as did Diogenes, and look for a man.

Your Xmas letter was most pleasurable.

Newman's "Loss and Gain" is rather hard reading, if one be looking for a story. Try "Fabiola." If you can enjoy Newman, his "Present Position of Catholics" will do you good. 'Tis very cleverly written; you will gain much by reading it and the "Apologia" slowly and between fishing hours.

God bless you all. Be sure to give me a great joy by telling me you have done my wish. Love to all.

W.D., S.J.

P.S. I am thinking of dear Gilbert's vow-day. He is praying for us now. Get Nat to join in praying for me.

To a Sister of Notre Dame:

Innsbruck, Universitäts Str. 8
19-1-'08.

Dear Sister,

Many thanks for your letter, the greetings, and especially for the prayers. I am in greater need than ever that you lift up hands and heart to the dear Lord who is so good to me. On the 2d of February I shall take my last vows. Seventeen and a half years have I been looking forward and longing for this day of utter sacrifice. To think of standing before the Christ of the Cross of glory, and of bearing for eternity the glory of the Cross upon me,—'tis a joy I could not and would not put to words. The glory of the Cross to the Jesuit is the glory of the solemn and irrevocable contract with the Catholic Church, which he takes oath to, on his last vow-day.

I hope you make out this scrawl. My fingers are cramped from over-much writing today. I cannot wait for the "uncramping." I am most desirous that you pray much for me; that you offer for me any and as many Communions as you can, before the great deed is done and done never to be undone by any power upon God's earth.

The *Record* came. It is very well written and edited. I read with special interest the paper on Mme. Swetchine. While en route from Marseilles to Beirut, I often spoke with the Princess Gagarin, niece of the great letter-writer, and wife of the Russian Consul-General of Beirut. The Princess was very well read. Her branch of the family is Schismatic, though her uncle was Fr. Gagarin, S.J.

Tommie is a great friend. He was one of my star-boys for the Sanctuary of St. Aloysius.

The death of Dr. Stafford [the celebrated preacher, pastor of St. Patrick's Church, Washington, D. C.] is very much of a loss to the Church. He was a man of such zeal and ability as to attract many whom the ordinary man could not stir. The people of St. Patrick's must be very sad.

Dr. Creagh's departure is indeed a loss to you. His influence must have been very helpful to the College.

Mother Loyola's books I know very well. Her "First Communion," is the best book of its sort I have read. I drew several yards of yarn from Mother Loyola's stock, when I spun yarns for the little ones of St. Aloysius. I hope my little Walter will enjoy the book.

I saw the Holy Father in private audience. We chatted about half an hour. I asked many, many blessings! *One for you!* I now send you this blessing with all my heart.

My stay here is of indefinite length. I am not likely to leave before July; and may be here next year. My studies are as interesting and time-absorbing as ever.

Now, dear Sister, pray for me all you can; ask the other Sisters kindly to help me in a like way.

May God bless you and keep you in the only life worth living,—the life of grace on grace, of virtue on virtue. Best love to Josie and the family, when you write.

Devotedly in Dno.,
Walter Drum, S.J.

Innsbruck, 12 Feb., '08.

Dear John,—

Yours of 20th Jan. is come. I am glad business is improving by degrees. Father Provincial will probably send you a document that I did my best to legalize properly. There is red-tape galore in Austria. I am pleased that Mary goes with Hugh to the Philippines . . . Do not so give yourself to the Charity work that your own work suffers. I like to see you helping in things Catholic.

I took my last vows on the 2d. The day was very happy indeed. I made a retreat of nine days before the great event. My time is my own. My work is (excepting an hour a week) done in private. I could readily give to the Lord the nine days of prayer. In the days of old, we used to beg from door to door, before we took our last vows. Times have changed. Such begging would not edify now; would be an occasion for a laugh or a joke, a cuss or an oath. Here in Innsbruck, the custom is to beg indoors. I went the rounds the day before my vows, and reaped a harvest well worth the while. At least forty Masses, and many prayers and Communions were offered for me. The Fathers and Scholastics were probably even more kind to me than they would have been to Austrians; they knew there would be a slight pang of sorrow due to the exile from my province. I have since learned from the States of the many prayers said for me there. God was very good. I made my sacrifice with joy. Though there was no pain in the sacrifice, it was an offering most acceptable to Our Lord,—even more acceptable for the joy of it, than would have been a sacrifice of sorrow. I offered the sacrifice first for the conversion of . . . . and his family and then for each of the family intentions; you all came in name by name and need by need, just as you have come into every Mass that I have said, just as you will come into every Mass God gives me grace to say.

Many theologians hold that these last Vows have such power of satisfaction as has Baptism, and remit all temporal punishment due to sin. What a great joy they are! When all the worry of time is over, and we are in an eternity of joy, without worry about family troubles, with never a thought of twenty nickels make a dollar, and such like, then one of our great joys will be my priesthood and my vows. When things seem somehow to go wrong, think how good the Lord has been to me and count my joy your joy.

My health has been remarkably good, though I cannot this year study so much as I did last year. The trips in Syria and Palestine and Egypt kept my stomach in fine trim; still I study as much as I could at Woodstock, and always have more to do than day-dreams and tooth-pick play.

When I do not write, 'tis for lack of aught to say. My life is uneventful. I could scarcely take it on myself to describe the Alps for three pages; so I just send a post card.

Lovingly,
Walter.

Innsbruck, Austria,
Passion Sunday, '08.

My dear Sister,

Thank you very much for your kind letter of 23d Febr., and still more for the prayers by which you help me to get that grace to will and do.

Your "penny a day" story is most amusing. Poor boy, I hope some Sunday he will hear the Gospel explained. Our Douai translation of the Bible was made when, I presume, the *penny* meant a *denarius*. An American version of the Gospel's would read truer "a quarter a day." The denarius was a silver coin, stamped with the image of the Roman Emperor, and used as a standard like the *franc* in France. If judged merely by its weight, this silver coin was worth not quite a franc. Our translation of a penny is most unfortunate. The Roman soldier of Our Lord's time received this coin a day,—it was the Roman poll-tax per annum. It was worth at least a dollar a day in purchase power. Some day I shall tell you of my dinner at Tarsus for four cents, of the staying power of one cent's worth of bread in Palestine, of the *fellahin* (day laborers) of Egypt who wear out their lives at eight cents a day. Dear me, dear me, how tedious all this must be! I came near to thinking I was preaching or teaching. Well, it may help you. I have a notion that our Sisters should teach such things to the children. The English Jesuits are now editing Scripture manuals for the schools. I fancy such manuals would give the Sisters just the information which they might teach the children. If our Catholics do not learn Bible bits from the Sisters, it is hard to see that the bits will ever be learned. The Sunday sermon is often exhortative and not at all instructive. But enough from the Biblist. Winter lingers still. So long as the Alps are snow-capped, and -cloaked, and -shoed, we may not hope for Spring and all its loveliness. I miss Woodstock Spring. The awakening of nature was ever an awakening of my fancy, a rest from the hard studies that are my life-work.

I have not been very well since last Xmas. The awakening of *dyspepsia* is my theme now. The old time headaches that so racked me at Woodstock, keep me from that study which I would like to do. Fiat!

I have not yet been on the Rhine. My way back to the States may take me by Mannheim.

My work for next year is still uncertain. I have not yet heard if I am to stay abroad.

I wish you a glorious Easter. May you rise with the dear Lord by a great increase of His grace. A Leo Logan of Boston, whose sister is at Trinity, is to be ordained here this year and has asked me to preach at his first Mass. Pray for me ever. Devotedly in Dno.,

W. Drum, S.J.

Innsbruck, 3 May, '08.

Dear Joe,—

Send on that letter from the Captain. Think of it,

I have not heard from him since I came to Tirol. There is a circus in town,—"The American Star,"—everything damnably out of the ordinary is *echt Americanisch*. It makes my hair stand on edge to think we are a lot of freaks and globe-trotters and nothing else. Well, I'll do some more trotting myself soon. I must loosen up my tightened brain, and broaden out my narrow mind. I wish I could broaden out my narrow waist. 'Tis all a waste. Even Tiroler beer fails to fill me out. I would I were more Keggish.

Now that you are taking sole-rest, try soul-rest, too. Give God a chance. The May devotions here in Tirol stir up many a man to a sense of his condition, startle many a man out of mortal sin. Even soldiers, who go to Mass only when the law of the land requires their presence but otherwise show little sign of religion, are present to hear the May sermon. The Blessed Mother often brings them back to grace. Tirol was once a very Catholic land. Even today, in Innsbruck, the traditions of Catholic Faith have not been crushed out of the people by socialism and anti-clericalism.

Well, God bless and care for you, Joe. On your 34th birthday I prayed for you earnestly. Congratulations.

Lovingly,

Walter.

Innsbruck, 3 May, '08.

Dear John,—

. . . I have sent you a few postcards from neighboring towns of Tirol. The Easter vacations last more than a month here, and gave me boon weather and companions for a few short excursions. My health is much improved by the exercise of these last days. Time has gone by quickly. My studies are my life. There is little of eventfulness. Day follows day in selfsame wise. Hence it comes that I do not write. There is nothing to write about, unless I take to word-painting. We are too prosaic a lot for word-painting. I think Joe is the only one of us who tried to be a *po't*.

In Feb. the *Stylus* published a letter from me and sent one copy . . .The *Messenger of the S. H.* for May published a pious little article by me. I have asked that you get a copy.

I have no word about next year. Our present semester goes on till the end of July. I attend no lectures but work in private. The great gain is that I speak German, and have no outside work to break in on my study.

The day I went to Kufstein, we five of the Maryland-

N. Y. province met Mgr. Marchetti, once auditor in Washington and now such for the German Nunciature. The diplomatic rules of the Papal government will not allow him to come to us, i.e., to pay a visit in another Nunciature,—without a special permission from Rome. Hence we met him on the frontiers of Bavaria.

The spring here is no spring. The Alps remain snow-decked and stare us out of all warmth. We have had snow within the past few days and are likely to have it again. The weather is not severe, but somewhat trying. The air is magnificent, bracing and clear. It is now raining and the atmosphere is not heavy as the East wind would make it to be in Boston. This year may be my last in Europe, and I wish to see much more before going home. I should shortly receive definite word from Father Provincial. At his last writing, he was thinking of giving me another year abroad; no decision has been come to.

You are with me in Mass and prayers. Am glad you are fairly well, and holding your place in business. Love to mother and to all.

Affectionately,
Walter.

Announcement of his brother's engagement reached him while he was on this journey through Austria and Bavaria, and the following letter brings his congratulations together with a full account of his trip:

Vienna, 28 May, '08.

My dear old John,—

I congratulate you with all my heart. I am as happy as can be over your joy. God grant that the great day, the *hochzeit* (high old time) as we in Dutch say, may be early in the fall. I have no word from Father Provincial as yet. We shall say, when we see what is to be. No need of any more words. You know me and I know you; and you know that your joy is more to heart to me than words can say.

I have your letters of 26 Apr. and 11 May. My postcards will have kept you in touch with me. I had had permission to hear a few prelections in Wien and München; but never had the plan to be so long from Innsbruck. I had to go. My nerves were on the jump. I was losing sleep every night. The Drum was unstrung; 'tis now well strung and in good tune. The constant strain of study, the Sirocco, the lack of recreation all worked ill for me. I am now quite fresh and frisky!

I bought a *rundreise* at the extravagant price of $15. Watch me and wonder at the route. I go third class, which is very cheap and dirty and all that. No matter, I go the round and meet the people. The poor are easier to reach than the rich, and fewer try to reach the poor than the rich. So, what if third class be dirt cheap and cheap dirt? I shall know more of the people and land them than do the starched and stiff tourists who go first class. The only trouble is that third class victims cannot generally take fast and through trains.

I was forced to wait about three hours at Salzburg, a fine old town that for centuries was ruled by prince-bishops. Thank God, our American bishops are priests and not princes! The old castle and many old churches were interesting. I went on to Linz and stayed there over night with our Fathers. We have the Third Probation in an old castle that was built by Archduke Ferdinand as part protection for the city; on completion, the castle was given to the Jesuits: it and its fellows were found to be behind the art of war. The castle is high on a noble hill, whence I got my first view of the lovely Donau (Danube). But the beautiful blue Danube is Missouri-yellow!

The morning of the 13th I left Vienna. My coupée companion was a painted horror, an old woman who had seen better days and better teeth and better hair. But in spite of all she spoke beautiful German, and that German was what I looked for. She was so selfish as to make me talk more English and French than German. But before the end I was laying down Catholic law in German. Like many others of the educated and impoverished people of Austria, she had lost the faith.

My stay in Wien lasted till the 21st. I heard three of the professors of the University; but spent most of my time in rest and sightseeing. Vienna is the most beautiful and most enjoyable city I have ever been in; and I have been in not a few. It has not so many boulevards as has Paris; but the uniformity of its buildings, the variety in architecture of churches and public edifices, the garden-streets, and the many little parks, the heartiness and cordiality and easy-going way of what the Germans call the *gemüthliches leben* make my stay here a very great joy indeed. It is a great thing to find this most Catholic city so well-organized and ordered.

The Stefanskirche and Votivkirche are the only churches of interest here. The first is a fine old Romanesque church with glorious intricacies of tracery and carving.

I was in raptures of joy over its old pulpit. How skilful the hand and painstaking the eyes that drove and directed chisel and hammer, while the intertwining foliage of that wondrous pulpit came into being. The Votivkircke is modern, pure Gothic, much like St. Patrick's in N.Y., rich in tracery work, marred only by a little paint. How painted stone worries me! The new church, called Canisiuskirche, is not as bad as the three older ones. It is a new Gothic, somewhat Arabic in form, with a pure and pleasing exterior but with an interior that does not please. I like severity, unless wealth be honest in its display. In the Byzantine of St. Mark's, Venice, there is a wealth of marbles and sculpture that I have seen nowhere else; the oriental display did not displease me, because it has truth and richness. What I do not like is untruth, painted stone and stucco work and imitation of marble. But I must cut my prattle short.

My address is as usual, Innsbruck. I entered the Lichtenstein Gallery by the usual scheme: Surpirse,—"I am an American!—I came 7000 miles to see this!" etc., etc. The picture galleries here are very fine, not to compare with those of Firense, but wonderful, and unique. It is good for my health that art gives me such joy. I should never be able to sit and semitise in season and out of season.

21 May, I went down the Donau to Budapest. The trip was very interesting between here and Pressburg, and in the vicinity of Buda; but for the most part dull and hot and wearisome. The Donau is nothing to be set in comparison with our Hudson and St. Mary's for beauty of scenery. The Potomac is wonderful in a grandeur that the Danube cannot have. For the most part this latter runs through low lying flats and is more like the Mississippi than our lovelier streams. The width of the Danube between Vienna and Buda is about three hundred yards. The depth is at times only about eight feet. I had an audience as usual and talked German to my heart's content. A German, a professor of Berlin, kept tagging after. He was loud in voice and strong in views, and drove his views home by poking me in my poorly clad ribs. He had been all the world over, and of course, took the wind from my sails many a time. With such a boor nearby, I could not give my usual speeches about the Catholic Church in America, the need of authority in religion, the folly of taking one's faith from newspapers and not from learned priests, etc. Still, I eluded my pursuer at times.

Budapest is very beautiful by night as one approaches on the Donau. The people speak Magyar, but I readily got on with my stock of languages, and found no need for Magyar. Our Fathers live in Pest. They treated me with the cordiality I received from the Jesuits the world over. Their church is small, but very influential for good among all classes. The city is beautiful, though not to be compared with Vienna. Only one church is interesting,—the old Mathiaskirche. It is fine old Gothic except where bad taste has improved it! The royal palace and museums are interesting, and such as I have seen elsewhere. What was of most interest to me was the Aquineum, a splendid ruins of a Roman legionary *castra*, a town, and the only temple of Mithras I have seen. I marked peculiarities that I had met no where else.

Another very fine feature of my stay in Budapest was the Hungarian music. A Mr. Beckett, an Englishman, whose son I have cared for spiritually in Innsbruck, kindly entertained me the last day of my stay and took me across the Donau to Buda to hear this gypsy music. The *zigeuner* (gypsy) plays by ear and renders the national melodies with a wildness and rhapsody that one cannot find in the Hungarian dances of Brahms and Jadasohn. I was charmed with the gypsy music. Mr. Beckett tells me that the gypsy father holds before his baby boy a fiddle and a florin. If the boy makes for the fiddle, the father turns him to music; if for the florin, the boy is brought up a thief.

25 May I started back by rail. Staid over night in Pressburg, and went out to Villa with our Philosophers the next day. The old cathedral is about the only thing of interest in Pressburg. The castle in which St. Elizabeth is said to have been born, is too much in renaissance style to be very old.

26 May at 3.30 p.m. I started for Vienna. I am resting much and studying very little. Here's a nice story for you. The day I went to Absam, I prayed especially at a famous Tyrolese shrine of Our Lady for your intention. Our Lady of Absam is special protectress of the marriage state the Tyrol folk go to the shrine in great numbers for the marriage ceremony. You see, Our Lady of Absam heard my prayer! It was your letter to answer my card from Absam that gave me the good news.

May Our dear Lord help you, John, and bring you at last that aid which you need. He has blessed you with good will and grace, and greater favors far than those that men of worldly ways want. Letter from Joe, including Hugh's to you, came today. Love to all. With my blessing,

Walter.

A letter from London gives the fact of his arrival there on August 16, and speaks of the arrangement by which he could reach Boston in time for his brother's wedding.

Passage was booked for him on the White Star liner *Adriatic*, sailing from Southampton. He had forwarded his portable altar in a special package, from Innsbruck to Southampton, intending as usual to celebrate Mass every day of the voyage across the Atlantic. The matter was given no more thought until, once on board, on the first day out, he bethought him of making arrangements for Mass the next morning. The valise was not there. He had not marked it "wanted," so he was told, and the valise was therefore placed in the hold with the other baggage.

> All day long the chief steward of 2d class, the baggage master, etc. made me fair promises and said me fair words, but gave me no altar. Even two dollars offered the room steward were ineffective. I went to sleep at midnight very sad. . . By ten the next morning, I realized that nothing had been done.

He was not going to submit tamely to the imposition that was being put upon him, and decided to bring the matter to higher officers. But even the Captain of the vessel would not at first grant the request that a search be made for the missing altar. But at last Father Drum won his way, and the boatswain was sent with four men down into the hold to investigate. The foul air of the deep hold, and the pitching of the vessel, made him quite sick, but he persevered in his search until he found the valise at 11 o'clock. At noon he was able to say Mass.

The following lines are culled from a letter written on the voyage, but not sent till a day or so after his arrival in New York.

> Written in good will, but forgot in the rush of many things.
>
> On Board R.M.S. Adriatic,
> 2 Sept. 1908.
>
> My dear John,—
>
> The trip has been rather rough, but not so rough as

to toss the Adriatic and myself too inconsiderately. I have been sick only part of the second day. The slight sickness was due to my going down into the hold to hunt my portable altar. The latter had been stowed away by mistake. After much fussing, I obtained it.

And in the same envelope a long letter, written after arrival gives further details of the trip across the Atlantic.

The College of St. Francis Xavier,
30 West 16th St., N. Y.

Dear John,

At last I have a little time to write up my sea trip. The crossing from Southampton to Cherbourg was not very bothersome. Worry and flurry were caused me by the absence of my altar. . . . We found the precious valise by 11 a.m. the following morning (the 27th Aug.). My joy was full, when by noon I was able to say Mass.

The joy of the Mass made me strong against the surging sea. I ate a good dinner and was not sick at all from that time.

At five p.m. we put out from Queenstown. The weather was rough most of the way, too much so for many passengers. I missed no meals and was in the very best of health and spirits. Many of the passengers were Catholics and I was very glad to be of help to them.

I had Mass every day at 5.30 a.m. after the second day out except for Sunday, when it was at 9.30 and on Thursday, when it was at 4.30 a.m. when we were off Sandy Hook. On week days, there were some twenty-five or thirty present in the sitting room for Mass. The sight was impressive and inspiring.

On Wednesday we had a terrific gale. It lashed the sea to fury. I stood out on deck and clung fast in seeming folly, but great joy. I have rarely had such pleasure in the sight of the battling elements, rarely seen so clearly and felt so tangibly the power of God in wind and wave. The looks of disgust on the faces of the fallen passengers were almost amusing, as I told them the storm had been the most enjoyable feature of my trip.

One thing to break in on the monotony of the sea was the many chances I had to talk foreign languages. A teacher of German in the High School of Portland. Ore., a dentist of Buenos Aires and his wife, a tapestry maker from some faubourg of Paris, a Syrian of the Lebanon, and others gave me the opportunity to brush away at German, Spanish, French, Italian, and Arabic.

The rest of this letter is concerned with the coming wedding of his brother John, and an attempt to solve the problem of getting up to Boston so as to celebrate the nuptial Mass, and not miss more of the classes at Woodstock, where he was already a few days overdue; and he also answers a difficulty that was then in the Boston papers about the famous case of Cardinal Casimir who became king of Poland.

CHAPTER XIII

PROFESSOR AT WOODSTOCK

In the beginning of September, 1908, Father Drum returned to Woodstock to take up his duties as Professor of Holy Scripture and Oriental Languages. "I fancy I am to become a fixture at Woodstock," he wrote to a friend who had inquired about his tenure of office. "For nine years I asked for foreign mission work; and even now it would be welcome; but along other lines are my activities permanently directed; and that by obedience." He succeeded Father John Corbett, S.J., who had taught Holy Scripture at Woodstock from 1905 to 1908; and was to hold the position for thirteen years, longer than any previous professor of this course. Of the twelve professors who had preceded him at Woodstock, the longest term was that of Father Piccirillo, S.J., who taught Sacred Scripture from 1878 to 1887, and that of Father Anthony J. Maas, S.J., from 1895 to 1904.

At first he found it extremely difficult to adapt himself to his new duties, because of his long absence from the class-room and the many distractions of his travels abroad. But with his usual energy he set to work to organize the course. For the first few years of teaching at Woodstock, there was a noticeable freshness and spontaneity about his manner and his method that made him interesting and popular to a degree. It was, therefore, most unfortunate, from the point of view of his work in the class-room, that in a very short time he began to undertake a tremendous amount of writing and accepted invitations for retreats and other extraneous labors that would have taxed the strength of a more robust constitution. His professorial duties were bound to suffer in consequence.

In the very nature of things, such a strain could not last long. The physical result was almost a foregone conclusion, and in the Spring of 1911, Father Drum succumbed to a nervus breakdown. A heavy cold set in about the same time and he was reduced to a very serious condition. Superiors accordingly decided to permit him to leave Woodstock for a complete rest, and he was told that he would not be expected to examine at the end of the school-year, or engage in any retreat work for the summer, until his health was completely restored.

At his own request Father Drum was allowed to go out to the Indian Missions in Montana for this convalescence, where he would be out in the open air a great deal and get plenty of exercise to recuperate his strength, and at the same time engage in some of the missionary work that would not overtax his weakened constitution. The permission for this vacation was granted all the more readily, as about this time there occurred some serious business eventualities in which the family were all concerned, and which would have required his presence at some time or other in the old home in California.

He spent about two weeks in San Francisco. The first Sunday of his stay he was asked to preach in the Church of St. Ignatius in that city, and his wonderful sermon on that occasion was printed in the San Francisco *Chronicle*, with the announcement that the distinguished visitor from the East was scheduled to preach again on the following Sunday evening. It is said that the crowd which filled the church on that day gave a collection ten times more than the ordinary. The *Chronicle*, July 24, printed an enthusiastic report which began as follows:

> Clear in his enunciation, a master of terse English, Rev. Walter Drum, S.J., delivered a strong sermon at St. Ignatius Church last evening, in which in no uncertain terms he spoke his mind on some of the questions of the day. He took a decided stand against the Socialism as preached by Karl Marx; he upheld the parish school as against the public school; and he condemned not the Protestant in his simple faith, but as he expressed it, the leaders of Protestantism who were preaching a Christless Christianity.

A few letters and some references from his later retreats give a detail here and there of his experiences on this trip to the West:

> St. Ignatius College,
> San Francisco, 12 July, '11.
>
> Dear little cousin,—
>
> Only two letters! That will never do! Be brave! follow my advice, and look to the little details whereof

I made mention. Go to Communion every week. Write to me! Be of good cheer. God is good, far and away beyond us in goodness!

I asked a little friend to send you four photos . . . and one for the D . . . to keep J . . . from the Y. M. C. A. I may not have told you that my Masses were for you all on the Feasts of Trinity Sunday, Corpus Christi, the Sacred Heart.

Westward Ho I am! The Vest is West and in the Vest am I! As we say here, "I blew in" last Sunday. The trip was slow, as I said Mass daily; sometimes a whole day had to be spent at such one-horse stands as Grand Junction and Las Vegas. More than a week went to both Al in Chicago and Hugh in Leavenworth. I was present when Hugh finished his year at the School of the Line. He ranked third and made honor graduate. Among the men he beat were chiefly West Pointers and two that had been instructors at West Point.

Mother is pretty well, and overjoyed to have me here. You will have to wait for a while and then I shall let you know more about my trip. Love to mother and each Koz.

Affectionately,
Walter Drum, S.J.

St. Paul's Mission,
Montana, 15 Aug. 1911.

My dear Koz,—

Here are your letters of ———. Many thanks, especially for the cheering report of the last. I wish you had made the retreat at Newport. It would have strengthened the will-power and cleared the reason-power and curbed the flesh-power . . . , and would not have worried you or unnerved you.

Tell John I am right proud of his manly Catholic spirit. God bless him! . . . My health is still a-mending. The steed I ride is Johnny,—not so lovely a name as Glory (my Woodstock charger's name). I ride a great deal, every day twice, or at least once.

"Welcome" by Mother Loyola, or "Under the Sanctuary Lamp" by Father O'Rourke would do for your Dominican friend.

God bless you, and keep your soul as I would have it. Be brave. Fight on! You will win! I won't let you lose! God wants you! Love to each and every one.

* * * *

I went out to Montana for a little rest on an Indian Mission,—far up near the Canadian border. The only good people nearby were the Indians. They were very good. In the Gros Ventre tribe of fifteen hundred Indians, there were only two marriages impossible in the Catholic sense of the word. The white towns, however, were an abomination. In a mining town called Landuskey, seven miles away, there never had been a natural death in about thirty years of the town's existence,—only homicides and suicides! While I was there, a baby died about five or six miles away from the town. The uncivilized miscreants, in cold blood, sent a protest to the parents of the baby against the happening of such a death in their neighborhood. I did not go to the town of Landuskey to preach. The priest in charge would not allow me. He said, he had gone there twice to say Mass; nobody had come. He had come to give a lecture on "The Greatest Mine in the World." Only two or three came; and they refused to stay, when they heard that the greatest mine was the Church. I begged him to advertise me as lecturing on "A Hell of a Mine", and was sure that they would come. Then I would turn on Hell hot and heavy, —as their doom, unless they attended to their Catholic duties. Many of them were Catholics; should have been Catholics.

Another mining town nearby was Lortmann, at the Ruby Mine, whose output was valued at eighty-five thousand dollars a month. It was a bullion, a mixture of silver and gold.

There I had to sleep on the floor. The good Jesuit missionary, who had charge of the place, had brought me about seven or eight miles to the town. That night, as we went along the route, the miners promised to come along to hear me. The Father lauded the speaker to the skies. The speaker gave his most fascinating grin. There were only two people there that night to hear my lecture! Two little children were true! I set them singing. Later we gathered in about twenty people, and I gave the lecture.

* * * *

My stay among the Indians was delightful [says a letter on the way to Woodstock]. Life among the Gros Ventres and Assiniboines agrees with me. Hunting and riding filled out much of my time. I preached several times, through an interpreter; and was a great favorite with Fat Captive, Jake Lame Crow, Dirty Shorty, Sits in the Middle, Mrs. Skunk, etc., etc. Too bad you could not eat some of the prairie chickens and sage hens that I shot.

To another man, the dangerous illness of the preceding Spring might well have been taken as a lesson in prudence, and induced him to curtail some of his many labors, and concentrate on class-work and Scripture studies at Woodstock. But it seems that Father Drum was of too active a nature,—or perhaps the explanation is that he still was subject to the dyspepsia that gave him such frequent headaches in the earlier student days, and that consequently his inability to stay long at his desk led him to seek for other outlets to his energies. Be that as it may, not many months went by before another illness attacked him, this time in a much more serious form. A heavy cold contracted in early December was quickly followed by double pneumonia. He was reduced to a dying condition, and received the Last Sacraments. But fortunately, and, as it seemed, almost miraculously, the crisis was safely passed, and the patient gradually recovered.

Perhaps this continued ill-health had something to do with the change that almost imperceptibly had come over Father Drum; he was, somehow, in the class-room, a different man. The vivaciousness and interest and spontaneity of former days gave place, for the most part to a formality, and a ponderous manner, that was altogether strange to those who had previously known him in and out of the class-room.

Many of his own brethren, .the younger among them especially, never had a chance to know Father Drum outside the lecture-room. He was generally too busy and far away to permit of any habitual contact, and too reserved to talk much about his own work, even on the few occasions when he would join a party around the walks on the Woodstock grounds. That he was generally an interesting companion, a ready and entertaining conversationalist—though always impersonal and gentlemanly, even when talking about his own experiences—all could recognize. They knew his wide reputation as a preacher and a lecturer and a scholar; but the man himself they knew only as a professor, a teacher. It might be remarked here that in the estimation of many, Father Drum, as is often the case with learned men, did not measure up to expectations in the class-room. Either he could not come down to the intellectual level of his students or he was unaware of the fact that he was often teaching over their heads. He himself admitted that somehow he was not having the same response from his classes as in the Scholastic days. One of his pupils, who was under him both in college and later as a Jesuit Scholastic, expressed to the present writer his astonishment at the change. In the old days, there was never a better teacher, one who not only knew his matter thoroughly but could present it in a lively and interesting manner, as well as with tremendous energy, while the results spoke for themselves at the time of examinations. His boys usually did exceptionally well. It was especially in Greek that he had excelled, and in this his pupils were thoroughly well grounded.

But later on, in the Society, things appeared different. It is hard to say what the trouble was, or why the apparently poor results. There was no shyness certainly, among his own; nor was it lack of preparation. Often, while speaking, he seemed to be feeling his way through a maze of possible expressions, speaking deliberately, ponderously, as though to make sure of the most correct nuance of phrasing and of meaning, the nicest accuracy of expression, so that there would be absolutely no danger of misunderstanding or misinterpreting the doctrine he was expounding.

An eminent Jesuit, whose authority as a writer and an educator is unquestioned, once said that in his opinion the matter and the manner of the Scripture class as conducted by Father Drum were most excellent, in accord with the best traditions of the Society, and with the prescriptions of the Ratio Studiorum, but that his classes were frequently unable to meet him on his own intellectual ground. The lectures were as a rule too learned for the majority. Father Drum would at first read through several passages from Scripture, and then proceed to a thorough examination of the text from the point of view of exegesis, the interpretations of the Fathers, the various theories of modern scholars and the decisions, if any, of the Biblical Commission, and occasionally showing the bearing of the text on important doctrines in dogmatic theology. As a rule, one felt that he was carrying prudence and orthodoxy to the very extreme of carefulness, and to this end all regard for the interest or attractiveness of the style was being sacrificed.

During recreation one day, the talk turned and the verdict was agreed to by everyone in the gathering: "You may say what you like about Father Drum's method and his class. For me, he is the *Roman Catholic*—; you can feel absolutely sure that he is always giving us thoroughly Catholic doctrine."

"I shall never forget," said a young Jesuit priest, "what he said when once I asked him a question on some point of doctrine, and his answer: 'If you had asked me that five years ago, I would have answered differently. But now I think that is a matter of faith, it is *de fide.* I have gotten to be antediluvian in my orthodoxy, and am not ashamed!' "

Quite frequently, however, he would lecture to the class from a few jotted notes, as was his method in preaching and in retreats. Then he was sure to be of absorbing interest, for he had a most fascinating manner of declamation. All undue formality would vanish, and one would wonder if it were really the same man that was speaking. Most secular friends and admirers saw him only in this one attitude, and judged him by this more attractive manner. He was decidedly more natural and at ease on the platform or in the pulpit than in the class-room. It was, indeed, the general opinion among the students that on those occasions when Father Drum branched off to relate some experience of his own or some anecdote of his travels, or better still, paused to elucidate a difficult point in dogmatic theology, the class was much more interested than in any other part of the Scripture hour. Many have testified that they found their difficulties in dogma clarified by Father Drum's ability to summarize and analyze an important thesis. They often expressed the wish that he would just let his notes go, and simply talk. However, he himself did not share this opinion, or seem conscious of the impression he was producing, as is clear from the following paragraph in a letter to a nun who had helped him with the typing of some of his lectures:

Woodstock, 14 Feb./14.

My dear Sister,—

Thanks for the neat and beautiful copy of the talks on education. It will be a temptation to read rather than talk the matter over next time. The difficulty is that things read do not impress to the full extent of things said. Here at Woodstock, for instance, how much easier it would be for me once and for all to write out my lectures; and thereafter to reel them off upon the screens of the fancies of my students! Yes, easier, but infinitely less effective than ex-tempore talk upon merely skeletonized jottings. Don't you think the like would be the case at retreats?

However, one must not think for a moment that Father Drum's sense of humor ever deserted him, or that he could not enliven a dull hour by the most unexpected quips and sallies. His was a droll sort of humor and quite original, at times startling, because no one but he could think of or carry off such a sally. It was of course a humor always gentlemanly and impersonal, but at the same time, quite exotic.

Someone raised his hand one day to ask a question as to the meaning of a certain mark in the Hebrew text. He was a big powerful man, and was not asking in fun. Father Drum did not answer at once. He looked down at the questioner, beamed at him a moment, then looking around, said in his slow, droll manner, "Birdie, wait a little longer, till your little wings grow stronger." There was no class for the next few moments.

"Eastern cities, and all ancient cities generally, are built on a hill—presumably for purpose of defense. But I rather think it was for convenient drainage." And as he spoke, he would look up and grin, showing white teeth and gleaming eyes.

Speaking of the Scriptural use of the words "heart," "bowels," etc., for the seat of the affections, he said once: "Love does not really proceed from the heart; but still, think of the small boy telling his mother: 'Mama, I love you with my whole brain!' We may have to come to that some day. The lover will have to talk by the card, and tell his beloved that he loves her with all the affection of the great sympathetic nerve that runs up his spinal column!"

"Sometimes even Homer nods," and Homer's pupils, too. When he noticed that some one was drowsing, Father Drum would look around a moment to attract the attention of the rest of the class, and then say in his deepest bass: "Awake, arise, or be forever fallen!"

During his first years as professor at Woodstock, Father Drum's knowledge of languages was manifested in a remarkable manner.

We have seen how in Syria he had within the first six months picked up enough Arabic to be able to travel about without the aid of a dragoman. At Woodstock he continued his language studies. By some unusual concurrence of circumstances, it happened that a very cosmopolitan community was gathered together at the college during those years. There were some Jesuit Scholastics to be found who spoke German well, others French or Spanish or Italian, and by great good fortune, there was a Scholastic in theology who had spent nearly nine years in Syria, and was able to converse fluently in Arabic; he belonged to the Province of Lyons, which is in charge of the Mission of Syria.

Father Drum was quick to take advantage of all these favorable circumstances, and soon had agreed to form a "language band" for conversational purposes during the evening recreation for every day in the week—a different language to be spoken each evening. But this was not all. He was speaking one day of the new fads that are now current in the teaching of modern languages. In the course of the conversation he let fall the remark: "I have no faith in these inductive methods. The old way of conning paradigms, and learning the essentials of grammar by rote is the only safe way. I have done it for *twenty-seven languages!*"

What was the full list of these twenty-seven languages? Very likely he included some branches or dialects of the same tongue. At any rate, the list certainly includes all the Biblical languages, like Hebrew, Syriac, Latin, Greek in its three periods, Classical, Hellenistic, and Modern; Arabic, Aramaic, Rabbinical, Babylonian, Coptic, Assyrian, and Samaritan; while among the modern European languages he could speak and write fluently in German, French, Spanish, and Italian. He had also a reading knowledge of Gaelic. The attempt to maintain his hold on all this complicated system of languages proved in the end too great a strain and useless burden. He gradually relinquished the effort, especially after his illness, when he continued to read and study only such languages as were necessary for his Scriptural work.

In connection with his work as professor, he had in course of time collected a valuable library of books on Holy Scripture, on the Eastern languages, on archeology, and such allied subjects. His room at Woodstock was fitted up for him with shelves along all the walls, reaching as high as the ceiling; and in the course of ten years or more he accumulated a varied assortment of volumes that were arranged in orderly fashion around the entire space. He thus lived in a library of his own. Nor was he ever at a loss to find any volume in all that bewildering collection, such was his systematic orderliness and method. As the writer had many occasions to see, when consulting Father Drum, he had never to lose a moment in finding a book on any subject.

In any large college or community, the question of books is sometimes a trying one. Important volumes may be misplaced and reference books may be removed from the library. A busy man, who needs all his material ready at hand, will feel keenly the waste of time and energy required to hunt for books that are not in their proper places. It was for this reason that Father Drum obtained permission to collect this special Scripture library for his own use. It was but natural, therefore, that he should be somewhat reserved in lending his books. On the surface, this may appear a selfish policy, but only the unreflecting will form such a judgment. Father Drum had suffered considerably from the neglectfulness of well-meaning but thoughtless people to whom he had lent books. Some of the volumes were allowed to stray or were lost altogether. Thus the good which he himself might have accomplished was not unfrequently hindered by these petty annoyances. It was, therefore, a necessary sternness, and a means of self-protection that made him so unwilling to allow his books to be removed. It is only fair to add, however, that when there was a genuine need, and it was possible for him to do a real service, Father Drum was most willing to permit others the use of his books. In one instance the present writer had occasion to consult several of his most valued volumes. Father Drum explained fully the reasons for his seeming rigidity on the matter, and then gave the books that were requested, with the words: "Here you are. I have been taught by sad experience; but I know you, and can follow you up if they are not returned in a reasonable time."

Father Drum's system of note-taking is a veritable object-lesson in efficiency and thoroughness, as well as an invaluable aid as a time-saver for a busy scholar. He adopted the plan at the very beginning of his Noviceship, adhering to it faithfully and industriously throughout his whole course in the Society.

It speaks well for his good judgment in the early days of the Noviceship, that he was able to select such a plan of note-taking that would be adaptable to every conceivable variety of subject that was likely to be useful in his later work. He followed a card system of his own, writing down all his data for retreats, sermons, etc., on separate small sheets of paper of a uniform size, about four by six inches. He used stiff cards of a slightly larger size, with raised headings, to separate the topics one from another. Later, at Woodstock, he had the carpenter construct for him five or six card cases, each about thirty inches in length and seven inches deep, to hold his ever-growing collection. All the boxes were quite filled. The notes were arranged as far as possible on the plan of the meditations in the book of the "Spiritual Exercises." Outside of those subjects, the material was filed according to an alphabetical system of his own, for which he kept a careful index. In the earlier days he wrote out his sermons on these small slips of paper. He took notes on every conceivable subject in connection with preaching, lecturing, class-work, and priestly ministrations in general. While on his travels in Syria, Palestine, and Egypt, he adhered to the same plan for his diary, which slipped into place in the same orderly fashion as did his notes on any other subject. However, during the period of his studies he deviated somewhat from this plan. This was because he wished to follow out that rule of the Scholastics which prescribes that each one should take notes during the classes and lectures, afterwards arranging these in a more orderly form in a blank book. This was but one instance of literal fidelity to the rule. In this fashion he filled twenty or more composition books, of the same large size, with notes on every thesis in philosophy. He later did the same for his theology. All this was done with a thoroughness and an exactness that is difficult to understand in one whose health, as we have seen, was not the most robust, and whose attention, too, was not occupied exclusively with these serious studies. We have noticed the amount of writing and of literary reading that he managed to accomplish during the same period. Perhaps the best explanation of this capacity for extensive as well as thorough work is to be found in the fact that he had early schooled himself to quick and efficient effort, and to an orderly arrangement of his time. Thus he was able immediately to concentrate his attention on the matter in hand, and to dispatch the work with minimum loss of time and expenditure of energy. It was well known that he could take a book of an evening, and having read through the entire volume in a few hours, prepare a written estimate of its contents and its value. And this not merely with lighter literature, but also with books on more serious subjects in his own particular line.

His duties at Woodstock also included, from 1909 to 1919, the care of the Woodstock College Library, which contains a wealth of important books. Under his care, there were many valuable acquisitions of Scriptural and Oriental works.

As happened everywhere else, the War brought serious financial difficulties to Woodstock. None understood the situation better than Father Drum, who usually managed to keep himself accurately informed on all that concerned the welfare of the Society. With his usual initiative he set about to do his share to help settle the general difficulties. After many plans had been considered and suggestions made by some of his friends, it was decided that the most efficient service could be rendered if he were to organize his own branch of the Woodstock Aid Association, which had been launched by the Province as a body early in the fall of 1919.

Briefly, the problem was, as Father Drum often took occasion to explain in sermons and letters, that in the Province there are nearly half the members of the Order who must be supported during their entire period of training, some fifteen years, by the voluntary contributions of the Faithful. They are supported with a minimum of expenditure a day for books, clothing, lodging, travel, and everything else. Yet even on this economical basis, in 1918, when there were 532 men to be supported in training, the sum of $195,180 for the year had to be raised. In 1919, with 553 men to support, the sum of $201,845 was needed. And so the total expenditure was bound to increase with the increase of candidates for the Novitiate even though the actual expenses were being kept down to the minimum. If the money could not be raised, the Provincial was faced with the disastrous alternative of turning away Novices for the Society; and yet the work of the Order was calling for more and more needed candidates each year.

Contrary to an unfounded impression, the Jesuits in America have no source of revenue from which to draw. In the Maryland-New York Province, there are some farm lands in Maryland that bring in a small revenue each year; the farms at Woodstock and St. Andrew-on-Hudson help a little towards the expense of both houses; and Woodstock has a slight foundation sum. For many years, a tax on our Colleges and High Schools, and the offerings given to the Fathers for their Parish Missions, all bore a noble share in the expenses of training Jesuits for the work of the Society. Now the Missionaries in the Philippine Islands and in the Island of Jamaica have to be taken care of by a fund that has no natural power of increase. The increase must come from some external source.

Happily, a solution was found to a certain extent, in the organization of what was at first called the Woodstock Aid Association, according to which individuals were asked to become members by the contribution of one dollar per year, or whatever they chose to offer over and above this sum. The Association was organized in the Jesuit churches and schools in the Province, and is still functioning, though under the new and more comprehensive name of the Jesuit Seminary Fund.

To those who knew Father Drum and the wide influence that he wielded, it was not surprising that he was able literally to found his own branch of the "Aid" as it was called, among his hosts of friends. It was given the name of the St. Ignatius Branch. In the fall of 1919, he had personally enrolled an army of 745 members actively working for the good cause, divided into bands under forty captains; in 1920 the number of members rose to 1088, under the leadership of fifty-five captains; and up to the close of the fiscal year in June 1921, his carefully kept records showed a membership of 1136 active workers for the "Aid," in charge of sixty-two captains.

When the returns began to come in, the labor of keeping the books became too much for Father Drum and interfered with his duties; so it soon became necessary to ask the help of the organization. Most of the office work was transferred to New York, where Mrs. S. D. Lorzer was made president of the branch, and Miss Irene Germain, secretary, while Father

Drum retained full supervision and control of the organization. Since that was a period of general hard times, it was found inexpedient to beg outright for money for the "Aid." Instead, scores of other legitimate ways of raising funds were resorted to. There was a Mite Campaign which covered the whole Province, the captains working in well-defined zones; there were euchres, raffles, bazaars, whist parties, and other social affairs; there was a campaign to interest talent of various kinds for the same good cause, the result of which was that a traveling gift-shop was established, where novelties of all descriptions were sold, "all showing what a band of loyal, whole-hearted workers could do, while trying to return in a material way a little of that devotion Father Drum gave so lavishly to all," so said the president of the St. Ignatius Branch. To give a few items of the practical results: The four months of routine work, from March to June, 1921, netted $650. On December 3 of that year, while Father Drum was ill at the hospital in Baltimore, a Christmastide sale at St. Francis Xaviers's College, New York, netted $700. The 1922 Mite Campaign from New York and Philadelphia alone netted $1,125. With his usual instinct for thorough business method, Father Drum had an annual report printed at the Woodstock Press, and the last fiscal year under his supervision, ending with June, 1921, his branch had brought in to the "Aid" a total of $14,487. Besides this sum, there were four burses for Woodstock College, one of which amounted to $7,082 (the sum required is $8,000), while the other burses were making rapid progress.

CHAPTER XIV

LECTURES AND CONFERENCES

In response to a request for information with regard to the lectures that were delivered by Father Drum before the Brooklyn Institute of Arts and Sciences, the director, Mr. Charles D. Atkins, forwarded the following list, and added a brief appreciative comment:

1914-1915 COURSE ON "THE POETRY OF ISRAEL"

Nov. 12, The Shepherd Song
Nov. 19, The Pilgrim Songs
Nov. 26, Temple Anthems and Exile Elegies
Dec. 3, The Essentials of Hebrew Poetry
Dec. 17, The Great Anthology of Hebrew Poetic Thought

1915-1916 COURSE ON "TWENTIETH CENTURY SEARCH AND RESEARCH FOR AN HISTORICAL CHRIST"

Nov. 11, The Liberal Christ
Nov. 18, Phantom Christs
Nov. 25, Eschatological Christs
Dec. 2, Mythic Christs
Dec. 9, The Traditional Christ

1916-1917 COURSE ON "THE IRON AGE OF ISRAEL"

March 29, Critical Theories,—Astruc to Wellhausen
April 5, Conservative Defense
April 19, The Stories of the Creation and the Fall
April 26, The Deluge, Babel, the Patriarchs
May 3, Abraham and Hammurabi; Sodom; Jacob; Joseph.

We desire to state that Father Drum was a very stimulating and effective lecturer who always attracted large audiences. He was a man of strong and invigorating personality, and of scholarly attainment. The Institute discontinued his series because it is accustomed to vary its lecture program from year to year and had offered about all the lectures Father Drum had available at the time.

Mr. Atkins wrote to Father Drum at the conclusion of the first lecture course in 1914, congratulating him on the unusually large attendance that had been attracted to his lectures, and inviting him to favor the Brooklyn Institute with another series for the following year. Father Drum saw in this second invitation an opportunity for a work of special usefulness which would enable him to put forward the doctrines of the Church before an audience that would otherwise be inaccessible to a Catholic preacher. "Last year, at the Institute, we spoke on Poetry; this year our subject will be Apologetics," were the words with which he began the first lecture of the series, and he resolved to go over the whole ground of twentieth-century criticism and research on the great question of the Divinity of Christ. The choice of subject was, indeed, masterful and attracted immediate attention.

Monsignor John L. Belford heralded the course of lectures in the Nativity *Mentor*, and urged all Catholics to attend them. The New York *Freeman's Journal* also spoke editorially of the great value of the course which Father Drum had given in 1914, and added a commendation of the liberal policy of the Brooklyn Institute in thus listing Catholic priests on its lecture program:

> The Brooklyn Institute is not a religious organization; and yet, it is only recently that priests have been invited to appear before its assemblages. The past year saw three priests listed in the Institute Bulletin to lecture. They were the Rev. Cornelius Clifford of the diocese of Newark, the Rev. Henry Brown, S.J., President of the Irish Classical Association, and Rev. Walter Drum, S.J. This year the bulletin lists only one priest. And yet even this is in contrast to the narrow policy of the Lowell Institute of Boston and the University Extension Lectures of Philadelphia.

If the first course of lectures in 1914 was exceptionally well received, the enthusiasm which greeted the series on the "Twentieth Century Christ" was even more remarkable. The hall was always crowded to overflowing. Here was a lecturer who combined extensive erudition with vivid dramatic ability and an attractive personality. He was a specialist on his subject. He had traveled extensively, and knew the languages of the Bible. It is, therefore, scarcely to be wondered at that the course was so largely attended by priests, ministers, rabbis, Sisters, college professors and a varied gathering of the educated laity. "Despite the heaviness of his subject," says the *Freeman's Journal,* "the speaker seemed always able to hold the attention of his hearers." He outlined the whole course of modern rationalism, showing how the great professors of non-Catholic universities in Germany, England and the United States had begun by rejecting all idea of the supernatural from the Gospel story and ended by reducing the Divinity of Christ to a mere shadow. Rischl, Bauer, Renan, Harnack, Loisy—he quoted from them all. He had read their works. The sincere thinker could not but be impressed with this array of proof and the exuberant erudition displayed by the lecturer. The traditional view of Christ as the Son of God was expounded in a manner at once clear and profound.

This course on the "Twentieth Century Christ" proved to be the most popular of the three series and Father Drum was finally induced, by some friends among the Knights of Columbus in Brooklyn, to put the entire argument together in a short pamphlet of thirty-six pages which he entitled "The Divinity of Christ," and published from the Woodstock Press. The first edition of 5,000 copies in 1917 was quickly exhausted and a second edition of 5,000 copies was printed, and sold before the year was over. A third edition of 10,000 copies, was nearly half sold out at the time of his death. The lecture course usually ended with a peroration that invariably lifted the audience to a pitch of tremendous enthusiasm. This finale was later embodied in the published pamphlet, and ran in part as follows:

> To sum up then, the Divine Ambassador of God the Father, Jesus the Christ, has consigned to a living body of teachers that message which He received from the Heavenly Father. This living body of teachers He has made to have two essential attributes and four visible marks,—it must be Infallible and Indefectible,—One, Holy, Catholic and Apostolic,—yea, Petrine.
>
> Now, find that body of teachers. It is Indefectible! It must exist today. Is there any teaching body today that dares claim these essential attributes, and these four marks, especially the last, the Petrine? . . . There is only one Church that has ever dared, or will ever dare claim to be such; and that is the Catholic teaching body, the great force that welds together into one faith seventeen million out of forty-five million in the United States, who admit to the census-taker that they have any religion whatsoever.

When I think of the great body of Protestant professors of Scripture, Theology and Philosophy, teaching their vagaries in Harvard with Lake, Royce and Hocking; in Yale with Bacon; in the Union Theological Seminary, with President Brown and the rest of the professors; in the Baptist Chicago University with Burton, Smith and Foster; way out in California University with Gayley and the Moravian Bade; when I note with horror that the very leaders of Protestantism, in our great non-Catholic Universities and important Protestant seminaries everywhere (with very few exceptions) in the United States, have given up the Divinity of Christ; when I think of all the Lutheran faculties in Germany, all the Calvinistic faculties in Switzerland, gone hopelessly away, gone boastfully and of set purpose away from the Divinity of Christ, when I see Oxford and Cambridge with Chairs of Divinity filled by men who deny the Divinity of Christ,—men like Canon Sanday, Lady Margaret Professor of Divinity in Oxford, and Canon Inge, Lady Margaret Professor of Divinity in Cambridge; and contrast these men with our Catholic professors, a picture occurs to my mind, which I saw in the National Gallery in London about ten years ago. It is entitled "The Unknown God" and is the work of a modern Götze.

The Christ is fast to a pillar, stripped, bleeding, crowned with thorns. On the pedestal of the pillar is the inscription *Ignoto Deo*—"To the Unknown God."

The allusion is to St. Paul's preaching on the Areopagus, the acropolis of Athens. There he told the Athenians, they believed in an Unknown God. From the worship of the Unknown God, Paul strove to lead the Athenians to believe in Jesus, the known God. Were St. Paul to come to the world today, the scene he might witness would be such as Götze has depicted.

The world passes by, and it has not a glance for the unknown God; it has washed its hands of the Christ, even as Pilate did long ago.

There is the sport, in his homespun tweed; a smile is on his face. He beams, as he reads the sporting extra. With his crop he slaps his putties, as he strolls along. He gloats over the victory of some favorite of the turf. He gleams; for he gleans by that victory! He does not look at the Christ. A horse is more to him than is the Christ! He has washed his hands of the Unknown God.

There is the lady in her satins. By her side is the dandy. She flirts with a lorgnette, he fillips a cigarette. He leers and lures and allures. She smiles and smirks and perks. Neither he nor she has a look for Christ. A lorgnette is more to her; a cigarette is more to him than is the Christ. They have washed their hands of the Unknown God.

Their goes the newsboy with his extra. There sits a forlorn, love-lorn woman, in rags and tatters, holding in her arms a sickly babe in less than rags and tatters. Neither has a look of trust or of faith in the Christ. Even they, that need him so, have washed their hands of the Unknown God.

And there struts the minister of the Gospel—the type of our Protestant University professors of Scripture—there he struts. In his hands is a great big Book. With all the dignity and self-importance of his class, he walks along. His head is buried in the Book. His eyes are all for the Hebrew roots of the Book. He has no look of faith and worship for the Christ! Why, a Hebrew root is more to him than is the Christ. He, too, the minister of the Gospel of Christ,—the professor of Scripture,—the educator of ministers of the Gospel, he too—God help us!—has washed his hands of the Unknown God.

Oh, I thank God with all my soul that I belong to a Church that is Indefectible, Infallible, tyrannical in its Infallibility and Indefectibility; and will never allow me to swerve one inch from the belief in Jesus, the Christ, very Man and very God, now and forever.

But the mere printed page can reproduce very little of the persuasive forcefulness and the living, earnest personality of the orator as he delivered this peroration. He usually abandoned his notes and stood out in front of the audience; he seemed to throw his whole being into the dramatic impersonation of each character as the picture was unfolded. On one occasion, the late Bishop Colton of Buffalo was moved to such enthusiasm by this outburst, as it was delivered in the hall of the Nardin Academy, that he rose from his seat and strode up to the platform to embrace the speaker, and then turned to call upon the audience to give a rousing cheer for the great champion of Catholic faith whom they had just heard.

This course on the Bible and the Divinity of Christ, proved to be the most popular of all Father Drum's series of lectures in the Brooklyn Institute. One of those who heard him and was as interested as any in the scholarly gathering that each week crowded the spacious hall, was the Rev. John T. Driscoll, of Fonda, N. Y., who a short time afterwards came to invite Father Drum to repeat the course at the next session of the Catholic Summer School. The invitation was accepted and the course was given at Cliff Haven, N. Y., in the summer of 1916, under the title of "The Bible and the Church." Here again the lecturer attracted considerable interest, and proved so popular that he was asked by the Catholic Women's League, of Boston, to repeat the course once again at St. Cecilia's Hall, in that city, under the auspices of His Eminence Cardinal O'Connell. The hall was crowded to its full capacity at each of the lectures.

This same lecture series Father Drum adapted to the form of conferences for Lent and Advent, which he gave many times over, under the general heading of "The Church and the Bible." From the many newspaper reports of these sermons and conferences, and from his own copious notes, in which the lectures are written out more or less completely, we can see that he kept to the same outline for the entire series, only varying his individual addresses to suit the exigencies of the occasion, and the needs of an audience that was generally far more sympathetic and uniform than the mixed audience which first greeted him in the Brooklyn Institute. There the attendance was largely non-Catholic and, at first, was to a certain extent hostile. As time went on Father Drum began to confine himself in these conferences to an exposition of the great historical proofs for the Divinity of Christ. "We need to go back to first beginnings, to fundamentals," he wrote under a clipping he had cut from an essay by Chesterton. In the beginning, he had chiefly occupied himself with a merciless exposure of the craft and unfairness of Luther's dealings with the Bible, and with violent denunciations of the modern un-Christian professors of divinity and of philosophy in non-Catholic universities.

> All this may seem to be too destructive [he said once in an earlier lecture], but the constructive element will come later. It is important first to show the utter weakness of the non-Catholic position, and afterwards to grasp the full strength of our own belief in the Divinity of Christ. But you say it is unfair—to single out the false teachings of professors in Germany, England and the United States. Why not those of Catholic countries also? Because there is no need of exposing them here. Priests

> who teach wrong are attended to by the authorities of the Church. But here I am listing the Christological vagaries that are tolerated and even taught by the leaders of the Protestant sects. No vagaries are *tolerated* in the Catholic Church. Not even a dangerous opinion is tolerated. There were Modernists for a time, under a mask. When they were exposed and condemned, they either submitted because they had not known the real trend of their ideas, —or they tore off the mask and became openly blasphemous, like Loisy, Tyrrell and Fogazzaro.

But perhaps the severest criticism came from those who had an exaggerated notion of charity and tolerance. "I never thought a priest would speak that way. That is most uncharitable!" was one such expression of opinion after his usual vigorous tirade against Luther and his mishandling of the Bible. Father Drum could not fail to take note of the criticism, and his next lecture began with a careful explanation of the distinction between good and sincere Protestants, whom he was not attacking at all; and the unscrupulous leaders whom they had blindly followed, and of these he did not hesitate to speak in the language of Our Lord Himself and of St. John the Baptist against the Pharisees.

> Charity is the love of God [he continued], and love of God cannot include love of the devil and his crew. St. Augustine tells us, where faith is not healthy there is no justice; for the just man lives by faith; he scouts the idea that they have love, or deserve consideration, who tear and lacerate the Spouse of Jesus Christ, the Church.

The course on the Bible and the Church was repeated again and again in many cities, and Father Drum's reputation was sufficient to attract immense crowds of people on every occasion. The Providence *Visitor*, after giving an enthusiastic review of one of these lectures, concluded its appreciation with these words: "Personally Father Drum is an ideal lecturer, in manner charming and magnetic, with a keen sense of humor, and delightful presence added to a mind extraordinarily gifted." Again and again the press reported the unusually large attendance that was attracted to hear him. "Father Drum gave the fourth of his Conferences on St. Paul to the large and enthusiastic audience of St. Paul's, Baltimore. As had been the case at all the lectures thus far given, the hall was crowded to its full capacity." For the Lenten course in the New York Cathedral, says the Philadelphia *Catholic Standard and Times*, "the seating capacity of the spacious edifice was overtaxed, and many hundreds stood in the aisles and vestibules." "I can still vividly remember the wonderful impression he made on my mind," said a Jesuit who heard him preach this course, "as he made a strong picture of the futile position of Protestants who rely solely upon the written word for which they have no guarantee. He held a limp Bible dangling in his hand, as he explained the necessity of having some expert and infallible guide to its meaning; and ended with an eloquent act of thanksgiving to God, that he was a Catholic and had a secure refuge from all doubt, in the authority of God Himself, who founded the Church of Christ to teach us all the truth." The Advent course in the Church of the Gesu, Philadelphia, was preached "to a congregation that on each occasion overflowed into the aisles and chapels." "The Richmond Course," he himself writes to a friend, "is playing to an increase of business. Last time there was standing room only." The course was also given in the Church of St. Martin, in Baltimore, for the Lent of 1916; in St. Mark's, Catonsville, the preceding year; in the Richmond Cathedral, in the Church of St. Ignatius Loyola, Baltimore, in the Wilmington Cathedral, in St. Francis Xavier's, N. Y., and in a number of other churches. The following outline is a sample of the general character of these conferences on the Bible and the Church, as they were given in the Church of St. Martin, Baltimore:

I. March 8.

SHIFTING SANDS

"A foolish man who built his house upon the sand." Mt. vii, 26.

Why the shifting sands of Protestantism? Because Protestantism is set upon the Bible. And "the Bible, the whole Bible and nothing but the Bible" means no Bible at all,—*nothing solid* upon which to set a religion. Why? Because in itself and by itself the Bible is no more than is the Quran of the Muhammedan, the Veda of the Hindu, the Sacred Book of any other form of religious belief.

The Bible is after the Church:

1. *In time.* Gospel and Letters of John were not written until about A.D. 110. Books of New Testa-

ment were not put together, until about A.D. 220. Meantime, where were the doctrines of Christ? There where he put them—in his Church.

2. *Apologetically.* Apologetics is the science of establishing Christianity. Christ could have established His religion in one of three ways:

(a) By the Old Testament way of miraculous intervention;
(b) By distributing an infallible Book;
(c) By an infallible Teacher.

He chose the last way.

3. *Dogmatically.* Dogmatics is the science of proving the doctrines of Christianity.

II. March 15.

ROCK OF AGES

"A wise man who built his house upon a rock." Mt. vii, 24.

The solid rock of the Church of Christ. How established?

1. Historical worth of the documents,—Mt., Mk., Lk., and Jo.—not as the Bible, not as inspired, but as human documents of that historical worth which Caesar, Livy, Polybius have.

2. From these documents, we prove that there was an historical person named Jesus of Nazareth; he said he was the Christ, and had a message from the Heavenly Father; he appealed to his future Resurrection in proof of the truth of his claim; he arose in fulfilment of that prophesy and in proof of his Messianic claim.

3. He gave this message to a teaching body, to have and to hold and to hand down to all peoples; he made that teaching body infallible, indefectible, one, holy, apostolic, Petrine, catholic.

4. Find that teaching body. It is indefectible. It must still be. There is only one teaching body that has ever dared to lay claim to have such prerogatives; and that is the Catholic Church.

III. March 22.

THE BOOK UPHELD UPON THE ROCK

"And the rain poured down, and the rivers rose, and the winds blew and beat upon that house; but it fell not, for it was founded upon a rock." Mt. vii, 25.

1. Upheld in its human make-up:
Style, words,—all that serves to preserve the Divine thought.

2. Upheld in its Divine make-up:
The two Testaments—the Jewish and the Christian Bible
The Old Testament:
The Canons of Palestine and Alexandria.
The Deuterocanonical Book—Judith, Tobias, Wisdom, Ecclesiasticus, Baruch, I and II Macchabees, together with parts of Daniel and Esther.
The New Testament:
Canon not infallibly declared by Church till Trent, A.D. 1546.

IV. March 29.

THE AUTHOR OF THE BOOK

1. The fact of Inspiration:
Protestant proofs:
Luther's—the inspiring Book;
Calvin's—the inspired reader;
Moderns—faith of the millions.
Catholic proof—the infallible Teacher.
2. Nature of Inspiration:
Protestant views.
Leo XIII, in *Providentissimus Deus*

V. April 5.

THE INFALLIBILITY OF THE BOOK

How far does Inspiration extend?
1. Not necessarily to the words and style;
2. To the thoughts. These are infallibly true in the sense intended by the Sacred Writer and by the Author of Holy Writ.

This inerrancy is a characteristic of every complete thought. Hence the Bible is infallible in its teaching of things of faith and morals, history of the human race, and all other things. And yet, though the Bible is a record of the history of God's revelation to man and of the human race in its handing down of this revelation, it is not a handbook of science. Therefore, the Bible does not professedly teach astronomy, geology, biology, etc., as it professedly teaches faith and morals and history. Therefore, it is unscientific to look into the Bible for information that one goes to scientific handbooks of natural sciences to acquire. Therefore, in the Bible, the phenomena of nature are spoken of in an unscientific way, in popular language, according to appearances and not according to rock-bottom scientific truth.

VI. April 12.

THE MEANING OF THE BOOK

1. Typical
2. Literal:
   Grammatical;
   Authentic and infallible.

Besides these lecture courses, Father Drum came to be largely in demand for special lectures on various occasions. It must have required tireless zeal and a vigorous physique to undertake the burden of so much public-speaking, and at the same time to lose never a day of class at Woodstock. Yet Father Drum seemed able to carry on, and in 1919 actually undertook to deliver three distinct courses of Lenten conferences in three different cities. There was the series of Sunday sermons on "The Church and the Bible," given at the High Mass each Sunday in the New York Cathedral; in the same Lenten season he gave the course of "Bible Talks," for every Wednesday evening in the Church of St. Aloysius, Washington, D. C.; and a third course on the general subject, "My God and My All," was given every Thursday afternoon in the Church of St. Ignatius Loyola, Baltimore. This effort would have been creditable enough for a man already burdened by his manifold occupations, even if Father Drum had given in all three places the same sermon each week. But the reported newspaper accounts indicate that the three courses were entirely distinct, and consequently he could not repeat himself. The following news item is taken from the New York *Catholic News* for April 12, 1919:

> Wednesday, April 9, the Rev. Walter Drum, S.J., of Woodstock College, gave the fourth of his Lenten conferences at St. Aloysius' Church, Washington, D. C. His subject was "The Solidity of the House Builded on the Rock." An essential part of this solidity is the inspiration of Holy Scripture. Next day he delivered the fourth Thursday afternoon lecture at St. Ignatius Loyola's, Baltimore, on the subject, "My Husbandman and I" and explained the Biblical allegory of the soul as a tilled field, wherein good seed and bad are sown. Sunday, April 6, Father Drum ended his course at the High Mass sermons at the New York City Cathedral. The subject of this last Lenten conference was "The Inerrancy of the Bible."

The following year a similar effort was duplicated in a course of Sunday sermons at the High Mass at St. Francis Xavier's Church, N. Y., on "The Rock of Ages," one of Sunday afternoon conferences at Marymount-on-Hudson, on "The Church and the Bible"; and a third course of Thursday afternoon conferences at St. Ignatius', Baltimore, on "The Mystic Christ-Life."

As was stated above, Father Drum was invited for a third time to appear before the Brooklyn Institute of Arts and Sciences, in the spring of 1917. His series of lectures this year were on the general subject of "The Iron Age of Israel," a title which covered his exhaustive study of the Mosaic origin of the Pentateuch, and his refutation of the "divisive theories" of non-Catholic Biblical scholars and of the so-called higher critics, from the time of Astruc to that of Wellhausen. The course ended with a vigorous defense of the traditional position of Catholic scholars on the various questions of the first five books of the Bible. The array of learning displayed by the lecturer, as in preceding years, attracted considerable attention and praise. Several professors from Columbia and elsewhere were present at the course, and one of these gentlemen took advantage of the permission accorded by the lecturer to propose questions and difficulties. The lectures sometimes ended with an open forum. Father Drum answered courteously and with his usual urbane manner, on the first few occasions but afterwards began to suspect an unfair animus in the questions that this particular gentleman was proposing. He gave the impression of speaking not for himself but as representing others. He always addressed the lecturer as Mr. Drum, never once according the least recognition to his clerical character. But Father Drum did not take notice of the slight. When, however, at the end of one of these lectures, this same gentleman rose to make the astounding statement that it was all very well for Mr. Drum to give the traditional view of the Biblical legislation; but that every scholar now knows that that legislation was derived from the Babylonian code of Hammurabi and was not at all of Divine origin, the lecturer could scarcely restrain his amazement and indignation. Evidently, the man's intention was simply to discredit his knowledge and ruin the good effect of the whole lecture.

Father Drum rose magnificently to the occasion. Drawing himself up to his full height, and pointing his finger menacingly at the speaker, he answered: "Now I understand; I have you now! You are not here to profit or to gain information by this lecture. Either you or those who sent you here have an unfair intention. But I will answer your objection. *I have read the Code of Hammurabi, in the original Babylonian.*" The he proceeded to compare, point by point, the two systems of legislation, showing how the few accidental similarities were such as might be expected because of similar conditions of race, geography, climate and economics; while the differences between the two systems were so radically essential that it was sheer ignorance to ascribe to them a common origin or to assert that either was derived from the other. The one code centered about the worship of the one true God, while the other code was wholly involved in the polytheism of paganism; the temporal prosperity and security of a privileged class was the whole aim and purpose of the Babylonian code, while internal sanctity and the equal application of the natural law to all men, under Divine precept, was the aim of the Mosaic legislation. These and other differences were pointed out and the answer of Father Drum to his objector was so overwhelming that the meeting broke up in prolonged applause.

However, it was evident that an atmosphere of strain and difficulty had been created in the course of these lectures. The director of the Institute found himself subjected to severe criticism, and there were open accusations that the Brooklyn Academy of Arts and Sciences was being used by Father Drum for proselytizing purposes. His name did not appear in the bulletin of lectures for the following year.

Speaking later, in a retreat, of the effect of these lectures, Father Drum said:

> Jews and Protestants complained that I was using the Academy of Music for proselytizing, and that, of course, was a crime fit for Blackwell's Island. I used to make, well, about ten converts a week at Blackwell's Island. There were not so many made in the Brooklyn Academy of Music, but there were some converts as a result of the courses of lectures. The lectures were simply Catholic truth, and rationalism might be allowed there or any-

where but Catholic truth was too unspeakable to be permitted.

With these lectures more than anything else, Father Drum may be said to have come into national recognition as an able exponent of Catholic Biblical exegesis. He came to be acknowledged as one of the foremost authorities on the Bible in this country. The recognition he received at first took the form of invitations to join various Biblical and Oriental Societies, and when, in the Spring of 1915, the plans were made for a Bible Congress at the Panama Pacific Exposition, the Secretary of the American Bible Society in charge of the Congress wrote the following letter to Father Drum inviting him to represent the Catholic position:

Reverend and dear Sir:—

Father Wynne of the Catholic Encyclopedia has suggested to me that I might write to you about the World's Bible Congress which is to be held at the Panama Pacific Exposition in San Francisco, August 1st to 4th.

The authorities of the Exposition having asked us to organize and conduct a Congress, we are planning the Congress on very broad lines as indicated in the telegram. Theological controversy at such a Congress would be out of place, and we will request the speakers to observe this simple rule, so that the points of difference between different churches will not emerge and no one who attends may feel embarrassed or wounded.

The main body of the papers is to be preceded by an historical Prologue intended to exhibit the Bible historically.

1. The Hebrew Bible as preserved by the Jewish Church, by a Christian Hebrew.
2. The Latin Vulgate as preserved by the Roman Catholic Church, by a Roman Catholic scholar of eminence.
3. The Greek Testament as preserved by the Greek Church, by a Greek scholar of eminence.
4. The Bible of the modern Protestant age, by a Protestant scholar and theologian.

As you will see from the printed slip our first thought, under Father Wynne's kindly suggestion, was to ask Cardinal Gasquet, as he is the head of the Papal Commission for the Revision of the Vulgate. His many duties and the conditions made by the war compelled him to decline, although he wrote a very friendly note. We are therefore turning to you, of whom we have heard in the Bible House, and whose reputation as a Biblical scholar learned in many things including the Latin Vulgate, makes us hope that you will do us the great honor and favor of preparing and delivering at San Francisco such a paper as will fitly represent the Church which you serve. We would of course pay all your expenses going and coming and entertain you while at San Francisco.

The passages marked in the telegram will show you that the Greek Patriarch of Constantinople has kindly agreed to send us a paper, though he will not be able to attend. The unity of Christendom, however marred by division, could be measurably exhibited in such a series of papers. We are well aware of our dependence upon the Roman Catholic Church for the manuscripts of the New Testament as well as for the Latin Vulgate, which is one of the priceless treasures of Christendom.

Can you not, even if busy, find time to prepare a paper not over half an hour in length, popular in form, on the Vulgate? It would greatly gratify not only the Society, but no doubt many members of your own Church who would go to hear you.

We are making the widest public announcements about the Congress as you will see. The extraordinary condition of the world at this time seems to invite all who call themselves Christians to hold up before the eyes of mankind Christ and His Word; and we would like to think that this Congress, which we did not originate but which Providentially seems to have been laid upon us, may be a rallying center for a new proclamation of the power of the Gospel. Will you not join with us in what we are trying to do, and let us have the benefit of your scholarship on this single theme which is so little familiar to many of our people and for that reason needing to be better made known?

On behalf of the Board of Managers,
I am
Yours very sincerely,
John Fox
Corresponding Secretary.

After mature consideration and consultation with the Provincial, it was decided that the good which might result from participation in such a congress was highly problematical at best, and accordingly Father Drum felt it necessary to decline the invitation, and did not attend the Congress.

CHAPTER XV

THE PREACHER

What was his method? Father Drum once told a friend that he memorized his first practice sermon, and broke down while preaching it. He never afterwards liked to memorize his sermons as written, and although he was for a long time careful to write out everything he intended to preach, he did not confine himself to what he had put down on paper. He found his manuscript too cramping to the rush of thought that usually came to him once he was warmed up to his subject. As time went on there was less and less time for writing, and, as he confessed, he came to follow more and more closely the precepts of Fathers Acquaviva, Polanco and other great authorities on preaching and to confine his preparations for a sermon or speech to fixing in his mind a clear idea of what he intended to say, and once in the pulpit or on the platform, to develop this outline point by point. He liked to look around his audience and single out someone—especially one who seemed inattentive or antagonistic, and throw the whole sermon at that individual—impersonally, of course. This, too, was his usual method in retreats. As a general rule he contented himself with a few jottings which served as leading ideas. A Religious recalls how he would spread out these small slips of paper over the table, like one dealing out cards, and then would talk without ever once looking at them. On one occasion, a Scholastic, wishing to prepare a set of sermons for the "Three Hours," on Good Friday, asked Father Drum for leave to look over his notes on these talks. It was shortly after Father Drum himself had given the "Three Hours" in the Immaculate, in Boston, in 1915. He turned to his note-boxes, drew out a few slips of paper and handed him one of them. "That is my sermon on the Third Word," he said. It consisted of just two texts from Holy Scripture and a sentence or two of his own.

Of course, such a method of preaching is difficult for an unpractised speaker, is often unsatisfactory, and is not usually recommended except after long experience; even in the case of a finished orator like Father Drum it is subject to the criticism, often made, that the style is rambling and the sermon lacks unity. In such extemporaneous speaking there is too often the tempation to abandon the projected outline and follow the tangent of a chance thought. Thus the whole sermon appears like patchwork. But this objection could not truthfully be urged against Father Drum's sermons for the reason that again and again he followed the same general plan once fixed upon, and he was as a rule careful to adhere closely to the guidance of his written outline. His usual method of procedure in a sermon was to begin with a thorough explanation of the Church's doctrine, often carefully and at great length, and sometimes in technical theological language that the average preacher would fear to use in a sermon, as being too learned for the majority of his hearers. Then there would follow some Scriptural or historical application, which often took the form of a narrative of some of his own travels, and gave to his discourse the interest of a vivid personal experience. In the peroration his favorite method was to describe some great and striking masterpiece of art, a painting or a sculpture, whose theme bore some analogy to the subject and gave a picturesque point to his lesson.

His own reverential thoughts on the dignity and importance of the preacher's office is evidenced in the following note which he had carefully written out:

> A preacher must speak well and not neglect elocution. The reverence that is due to the word of God demands this of him. He must avoid, however, too studied an elegance of style, lest the ear of his audience should stop short at mere words and eloquence, which would hinder the whole fruit of the sermon. He would thus preach himself and not Jesus Christ. When he has acquired a good style, his whole attention must be directed to this one object, viz., that grace may enliven that which art and nature have formed, and that the spirit of God may reign in his whole discourse as the soul animates the body. He must neither love, commend, nor value anything save what belongs to Jesus Christ.

It was very noticeable that he tried consistently to model his preaching on the style of Father Pardow, whom he greatly admired. Here and there in his notes, especially in those on retreats and conferences to priests, there are pages of hints that he had gleaned from his Master of Tertians. The latter's method is thus succinctly described by Justine Ward in "William Pardow, S.J.," in Father Pardow's own words:

> A sermon that follows a closely connected line of argument requires close attention and will not reach nearly as many people as a sermon that suggests a number of thoughts in a bold, striking way. We make the great mistake of considering our audiences as homogeneous like a class of theology. They are by no means so.

As regards manner and delivery, it is a remarkable fact that Father Drum appeared to be an entirely changed man in the pulpit. In private conversation, especially with strangers or mere acquaintances, he was usually reserved, slow and deliberate of manner, and gave the impression of one who weighed each word and was studying its effect on his hearer. No one could feel that he was an ordinary, casual individual. There was always the appearance of distinction, of pose, of the grand air, although it was not a manner that he seemed conscious of, or, once he was understood, gave any real offence.

But in the pulpit, he was a different man. He was more himself, more natural, full of life and interest. He used his voice with purpose and art. His training in elocution and his long experience with the management of dramatic productions, and his own ability to dramatize and impersonate character were all directed to one grand effect, and were essential elements of his success as a preacher. Some have said that his preaching did not conform to present-day ideals because it was too dramatic; and that Father Drum really belonged to the old school of expression. Be that as it may, the usual impression produced on his audience was that of interest and sympathy, though not always. He was too bold and uncompromising to please everybody. But there was generally an effect of sureness in the man, an impression of poise and reflection. It was interesting to watch him begin. As he folded his arms and looked quietly down and prepared to speak, his fine presence inspired confidence and expectancy. There was no hurry or excitement; the gestures were usually so deliberate and well-timed, you could watch them as they spread. Consciously or not, he seldom raised both arms at once, but most often employed the single arm gesture, from the breast out in a broad sweep, now one hand, now the other; and often the gesture would be sustained for a considerable time together. But best of all was the upward sweep, when his whole frame seemed to rise, too, as though wishing to draw the eyes and minds of the people upward with him.

Naturally he was impatient of hasty and offhand criticism; but the real humility of the preacher was evinced on many occasions. He would often ask for a frank judgment of his sermon, of his manner of delivery and the means of securing better effect. The following is part of a letter in which he discusses the good and bad points of his delivery with a friend, who was entirely disappointed with one of his sermons and had not hesitated to say so:

> Your critique was very acceptable; it tallied with my own adverse estimate. Father N. thought the voice muffled in the lower register. I deemed, with you, that the highest tones, which ran up because of my not feeling just myself, were the objectionable. Father Rector was pleased, said it was representative and apt. The indistinctness you noted in the higher tones, was, I think, due to too great speed of delivery, when my feelings got the better of me, the voice went up and the words were hurled out too speedily, a natural interference of tones resulted. The Gesu is a church that one must know in order to preach therein.

Many who heard him preach, even of his own brethren, could not understand the personal note that was so insistent in all his sermons. It seemed so much like mere egotism and vain show. But the simplest explanation of the matter is that he was himself really unconscious of any effort at self-display. He so merged himself in his subject that he had no hesitation to use all his material in what he thought was the most effective manner. He was an authority in his field, recognized as such wherever he went; he was a first-hand witness of many of the things he preached. As he had once said of a far greater preacher, the St. Paul whom he so loved and admired, no one can deny that he, too,

> had a message to give his audience; he believed fully in that message, and in his authority to give it; he presented his matter in such appropriate phrase and phase as to hold the attention of the audience, to grip the hearers unto his own

magnetic personality, and to swing and sway them the way of his own emotions and will. To do that is the chief element of oratory.

Father Hanselman, his former Provincial, once said he was always glad to hear him in the pulpit because of this convincing manner he had. "So and so said this. That is not true! That statement is false! I was there! I traveled to the spot, and I saw that scene with my own eyes!" This was the personal element which to many appeared so much vanity. But there is no gainsaying the value of a first-hand authoritative witness in any cause, and, once we are certain of the genuine sincerity of our witness, we can surely overlook any appearance of display.

His friend and admirer of many years, the Rev. F. E. Craig, of Mt. Washington, Baltimore, once insisted with much enthusiasm on the really wonderful impression that Father Drum's preaching always made on people.

> We could often get him out here, on some big occasion, to be Deacon to the Cardinal or in some other capacity, and he would be the preacher of the day. Invariably he would win the congregation at once and could always be counted on for a magnificent sermon. He had such an attractive face, he seemed to speak as much with his eyes and expression as with his voice, some even said that his hair and general appearance and manner enthralled his hearers even more than what he said. [It must, however, be added that this form of appreciation was always extremely painful to Father Drum himself.] But in meeting people, he seemed to have a way about him, an air of urbane courtesy, of ready wit and understanding that easily won him friends. Most people who once met him wanted to continue his acquaintance.

Whatever be the real cause of the attraction, his preaching seemed to have a peculiar fascination for the generality of audiences. Witness the crowds that filled the New York Cathedral for his Lenten Conferences, the Cathedral of Wilmington for the Knights of Columbus mission, the Churches of St. Ignatius Loyola, St. Paul and St. Martin, in Baltimore, for the Lenten and the Advent courses, and wherever he was scheduled to appear.

> The Church was crowded with men [announces a Richmond paper]. Father Drum, who is conducting the Retreat, is a wonderful speaker. He held his audience dur-

ing the entire discourse. His topics are entirely modern and a treat for non-Catholics to whom he extends a hearty invitation.

Despite the inclement weather, the attendance at the Mission being conducted by the Rev. Walter Drum, S. J., at the Richmond Cathedral, under the auspices of the Santa Maria Council of the Knights of Columbus, was greater than that of the opening services of Sunday evening. The course of lectures being conducted by Father Drum is unique, in that it is an appeal to the reasoning of all men, irrespective of creed.

The popularity [a Washington paper announces] of the Wednesday evening sermons by the Rev. Walter Drum, S. J., of Woodstock, is increasing, judging by the great crowds that are filling the large Church [St. Aloysius] each Wednesday evening.

Wednesday, the crowds were even greater than ever at St. Patrick's. Many non-Catholics attended to hear the sermon of Father Drum, on "The Laws of the Church."

The 286th Anniversary of the landing of the Maryland Pilgrims was called in the auditorium of Loyola College [1920]. Rev. Walter Drum, S.J., the distinguished Jesuit orator, gave a very complete lecture on Maryland history and the part taken in it by Catholics. The lecture of the noted priest was worth going a long way to hear. The applause and enthusiasm displayed by the very large audience present shows how appreciative they were of his eloquent address.

It would seem that his popularity as a preacher was acquired not slowly and by degrees, but almost from the very beginning, after his return from the Holy Land. There was scarcely an occasion when his appearance in the pulpit or on the platform was not matter for thoughtful and appreciative comment in the religious press, and often, too, in secular newspapers. To avoid the danger of misquotation to which the preacher is open, even more than any other public speaker, he took a useful hint from Father Pardow, and always made sure to have a written abstract of his sermon ready for the reporters. Father Pardow had found a great deal of trouble and worry through the faulty reporting of some of his orations.

What, then, was the source of Father Drum's power over a general audience? Everyone could recognize the full command that he possessed over all the arts of elocution and of delivery, together with vividness and variety of style, and an unusual felicity and force of expression. All, too, could acknowledge that his learning was far beyond the average. But more than that, Father Drum's sermons and conferences reveal, if anything, the deep living faith, and the genuine conviction of a man who had meditated deeply, and reasoned out for himself, and prayed earnestly over his faith—a man who had made every effort to saturate himself with all the available information necessary to give an account of that faith to the most learned as well as to the ignorant. There was no question of his absolute sincerity, and his knowledge of that whereof he spoke. As he often and often insisted, ours is a religion of the reason, not of the emotions:

> We should boldly take pride in our faith, and pity with genuine pity the benighted souls outside the Church. We, and we only, are God's people. Our faith is of the reason, of man's noblest faculty. Protestant faith is of the emotions; ours is "right reason's worship."

Hence it came about that his audiences were invariably impressed by the depth of his sincerity and by his tremendous earnestness, and deeply moved by his eloquence.

He was not, however, by any means the "cold intellectual" that some, judging by first impressions, would have made him out to be. He felt most intensely and deeply. Frequently an almost exaggerated emotionalism proved to be the most powerful element in his appeal. "The one thing I remember," said a friend who heard him preach the "Three Hours" in the Immaculate Conception Church, Boston, "is how he wept unrestrainedly at several parts of his descriptions of the Sacred Passion. He seemed to be living in and actually taking part in the scenes he described, and the congregation could not but follow with absorbing interest and devotion." It is true, however, that there was for the most part an air of dignity and hauteur, that betokened the scholar conscious of his knowledge and certain of his power; and although at times his manner took the form of a somewhat exaggerated effort at theatrical effect, the impression was usually forgotten in the irresistible rush of the preacher's feeling and, as it seemed to the generality of his hearers, the sheer vastness of his learning.

We may take it that he drew much of his power from the

St. Paul whom he so much admired, and in some limited sense, apply to him the description he often gave of the Apostle's preaching:

> There was no mincing of phrases, no glossing over of the great truths of his message, no effort at word-painting, nothing self-conscious; all is straightforward, direct, sincere, genuine, from the heart to the heart. Why? Because St. Paul was thoroughly honest, frank even unto brusqueness, free from all human respect.

As was said in a former connection, Father Drum once admitted that he loved and studied St. Paul with a veritable passion; and that, among all the Fathers of the Church, he admired especially the works of St. Jerome, who best interprets and illustrates the courage and the intrepid faith of St. Paul, and St. John Chrysostom, who best understood St. Paul's tenderness of heart and his charity. He once took occasion to expand this thought into a beautiful essay entitled, "The Heart of Paul," from which the above quotation is taken, and which was published in *Sursum Corda,* a magazine of devotional and academic literature circulating among the Confraternity of St. Gabriel. The essay ended with a tender description, under an assumed name, of his own visit to the great "Pauline" Pope Pius X. He said on another occasion, in one of the opening sermons of the course on the Phases of Pauline Piety:

> Our study of the piety of St. Paul has been limited. We have studied only one or two phases of that piety as evidenced by only a very few phrases of the great Apostle. In the full study of the piety of St. Paul there is a whole life-work. Every phrase and every phase of his life teaches us to know him better and better; to get down into his great heart deeper and deeper; to see for ourselves clearer and clearer how very much of Paul's heart is the Kingdom of Christ, how very little, less than little, of that heart is the kingdom of self.
>
> For "the heart of Paul was the heart of Christ," says St. Chrysostom. And if, by the side of that great and unselfish heart, we set our own small and selfish hearts, we shall see more clearly than before how great is the kingdom of self within us, and how little, sometimes even less than little, is the Kingdom of Christ in those recreant hearts which we have consecrated to the Sacred Heart forever.

At a later time he took occasion to make the further study

he so desired of the "Piety of St. Paul in Phase and Phrase." It appears that his ultimate purpose was to write a volume on this to him most attractive subject. What he did accomplish was to gather together a series of sermons for a course of conferences, which were first delivered in the Church of St. Paul's in Baltimore, in the Lent of 1917. The course proved to be immensely popular. He divided his study into several "phases" as he called them, of the piety of the great Apostle, as manifested in some of the most striking "phrases" of the Pauline epistles. Thus the "Ring Phase" was taken from 1 Cor. ix, 26, "I so fight, not as one beating the air," and was developed from this and similar passages where St. Paul uses the simile of the boxing-match to illustrate the spiritual combat of the Christian soul. Father Drum amplified the idea with great felicity by means of his own translations from the original Hellenistic of the Epistles. In a similar manner he developed the idea of the "Race Phase," from Heb. xii, 1-12, and from 2 Tim. ii, 5, etc. The conclusion of this conference may be of interest:

> As racers in the race of faith, we shall hear the *jeer* and the *cheer*. The *jeer* will be from the world; the *cheer* will be from Christ and those like to Christ. The world hated Him, the world will hate you; let it hate. The world scoffed at Him, the world will scoff at you; let it scoff. The world scorned Him, shouted at Him in derision, the world will scorn you, shout at you in derision; let it scorn, let it shout. The race will soon be over. He will be the Judge. The hating, scoffing, scorning, shouting world will be hated and scoffed and scorned and shouted at for all eternity.
>
> The *cheer* is from the Christ and the Christly, from the Blessed Mother, our Guardian Angel, our Patron Saints. In that encouraging cheer, there is one voice that sounds forth loud and clear, one voice that rings with the spirit of Christ. It is the voice of Paul. I hear him now. His words are from the heart of Paul. "The heart of Paul is the heart of Christ." The words of Paul are in the spirit of Christ. They cheered the early Christians on. Ringing down the centuries they cheer the saints of ages on. They cheer me on. They will cheer you on. That cheer of Paul, that cry of enthusiasm and encouragement to the racer is this; "Jesus Christ yesterday, Jesus Christ today, Jesus Christ forever."

The concluding paragraph of the "Ring Phase" ran as follows:

> The Pauline boxing-match between flesh and spirit is on today in every soul. And in many a Christian, the flesh is the victor. Right reason is superannuated. Strong will is retired on a pension. The emotions are supreme. Reason is blinded by the feelings; and blinded reason fails to direct the will God-ward; and a God-ward will fails to hinder the emotions in their pellmell, onward rush away from God. Legislation against white slavers does some good. The hounding of grafters now and then results in their reform. But the only effective reform of the state is the reform of each individual by his own God-aided efforts. The knock-out blow must be given by spirit to flesh. Self-conquest and self-restraint must be pratised to such an extent as to make reason and will supreme over the likes and the dislikes. Then and then only will grace-power be supreme in the individual; then and then only will the dynamic of Jesus Christ flow unimpeded through the Christian state.

In much the same way was developed the "Root Phase," from 1 Cor. iii, 9 and other passages; the "Building Phase," from Ephes ii, 21 etc.; the "Love Phase," from Romans viii, 31, whose most striking feature was the beautiful exposition of St. Paul's life in terms of an equation, thus: The personal equation was at first simply—Saul=Saul; but after the change, the equation was, Saul+Love=Paul. The "War Phase" was another effective study:

> St. Paul, while a prisoner, had often noted the armor of the pretorian guard. Every piece of that armor suggested to him the arms and defense of the Christian soldier against his foe. The girdle is truth; a cuirass of love and grace protects the heart; the shield of faith wards off the darts of infidelity; the helmet of hope prevents discouragement; the sword of prayer cleaves the way; the feet are ever shod with readiness to go the full way of the Gospel of Jesus Christ.

But the preacher rose to heights of lyric enthusiasm in describing the "Mystic Phase," the idea of the Mystic Christ so prominent throughout the Epistles of St. Paul. This was the chapter that seemed to grip most forcibly and hold his hearers enthralled with the beauty and the devotion of his exposition on the subject. Of its very nature, this discourse was most adaptable for retreats, and Father Drum did not fail to make effective use of it as a touching meditation.

There was a singular pathos in the fact that this discourse on the Mystic Christ was the last that he gave, to the Nuns of the Visitation in Baltimore, during the very week that he was stricken with his fatal illness. There had been difficulty in securing a priest to give the Sisters a retreat in preparation for the renewal of vows. Father Drum heard of the difficulty only when it was too late to arrange for a full Triduum, and asked if they would like to have the meditations of the Triduum all on one day. They were delighted with the offer, and made what Father Drum called a *uniduum*. The subject of the Mystic Christ was developed with even more tenderness than he had done in his public exposition of this theme. In St. Paul's Church, Baltimore, and elsewhere, he had begun with an analysis of the false mysticism that was being made so much of in modern rationalistic literature, and by such writers as Sir Oliver Lodge, Sir Arthur Conan Doyle and others, all of which he characterized with his usual vigor, as "a modern, up-to-date recrudescence of damnable necromancy, where it was not mere twaddle and tomfoolery."

But on this occasion he forgot his usual denunciations of false doctrine and was rather filled with the deep spirituality of the theme of the Mystic Christ, possessing a Mystic Head, Mystic Body and Mystic Veins. The Faithful form the Mystic Body; they are the members of the Mystic Christ; they are all one in Christ; hence the reverence we must have for God's commandment to love one another; whatsoever we do to our neighbor is in reality done to Him. The Sacraments are the veins of the Mystic Christ, carrying to all the members of the Mystic Blood, that is grace. As in the physical blood there are red corpuscles and white corpuscles, the former to carry oxygen and convey nourishment to the members, the latter to fight the germs of disease and decay and to preserve the health of the body, so in the Mystic Blood of the Mystic Christ are found the red corpuscles of sanctifying grace that give life and nourishment to the soul, and the white corpuscles of actual grace to fight against temptation.

It may well be said that in his preaching he was possessed by the spirit of his favorite champions of the ancient Church,

St. Jerome and St. Chrysostom. Some one has remarked that his absorption in the study of these Fathers, with their vigorous Catholicity and uncompromising attacks on heretics and the enemies of the Faith of Christ, surrounded him with an atmosphere of polemics that made it impossible for him to adapt himself to the milder methods of more modern controversial warfare. "The one feature of the retreat at Wilmington, which I am sure all the men would recall, was the militant Catholicity of the preacher," writes a priest of that diocese. As in his lectures and conferences on the Bible and his attacks on Luther and the Protestant position, he was criticized time and again for being too violent in some of his sermons, and urged by his friends to moderate his invective against non-Catholic scholars. For the most part, they argued, such men would not pay attention to any attacks from a Catholic quarter, and the effort might better be steered to a more constructive purpose.

There is a deal of truth in this observation. We all deprecate violence where gentler means are available and effective. Father Drum himself felt, after an unusually severe storm of criticism that followed one of his lecture courses, that it would be well to measure his blows in future. But everyone will be ready to allow that with some of our esteemed adversaries, a shock is absolutely essential, if we are to rouse them from their smug self-complacency, and convince them that their self-esteem and dignity are not beyond reach. It is encouraging for us to know that the Church will always find men of strength to rise to the occasion, and, unassailable in their own true lives and in the sincerity of their beliefs, to fight strongly, and hit hard, straight from the shoulder and at a very real enemy. "St. Paul is forcible and graphic here as elsewhere," Father Drum said in the sermon on the "Ring Phase," "his purpose in the ring, in the boxing match of life is really to *bruise* his enemy; he sets himself down as an earnest and an effective *bruiser* in the spiritual ring." That was himself. He was firmly convinced that there was real need of a show of strength upon occasion, in these days of laissez-faire and of easy-going compromise. The Church is not safe and God's work is not done, if we are all to adopt an attitude of perpetual apology and let our principles crumble before the indifference of a disdainful world. We are glad to see a warrior now and then arise to champion our principles and give us greater security in our own convictions.

Some people become nervous at the sight of a battleship, with its bristling guns and tremendous armament. It is so sinister, so violent, and they would have us believe, so unnecessary. Yet all the while, we are proud of this bulwark of our country, and feel safe.

Moreover, it must be noted that the *odium theologicum* of a great preacher and writer may at times be very bitter and very violent, and yet not necessarily personal. Father Drum repeatedly insisted on his one desire to stand by the traditional doctrines of the Church, the old war-cry that was given him by Pius X himself, and to safeguard the faith of the multitude who might be led astray by the false or incautious teachings of prominent men, and this aim he pursued regardless of the personal feelings of the teachers themselves. True, it is difficult to pursue this policy for any length of time and not create bitterness, precisely because it is difficult to point out an error in one who professes to instruct others, or a mistake in the utterances of a public character, without at the same time creating the impression that the error is due to deliberate intention, and perhaps to malice. We expect our teachers to be exact and careful in their instructions, and our public characters to mean what they say; and to point out their mistakes is often equivalent to criticizing their intentions. The one may easily be pardoned; but the other makes enemies.

Even in private life, one was sure of Father Drum's absolute respect, if one were sincere, loyal, single-minded. But he had no time, and when the occasion called for it, no mercy for insincerity, or for the mistakes of those who should have known better, or for falsehood coming from a man of influence.

"Overmastering faith, persuasive zeal, tender charity, these must be the characteristics of the preacher, the ambassador of Christ. He must be mighty in word and work, like his Divine Master, before God and all the people." Such in brief is the characterization that Millet-Byrne gives of the true preacher-priest of God; and those who time and again listened to Father Drum's impassioned eloquence will scarcely quarrel with the statement that he at least aimed at this ideal. In noting the effects of his preaching in our purely human way, since God alone, the Author of grace and of conversion, knows the full effects of the preacher's efforts, we can venture to hope that his words about the preaching of the eloquent Bishop Currier might be equally descriptive of his own success:

> Thousands are the souls [he said in the funeral oration preached in the Church of St. Ignatius Loyola, Baltimore] that were illumined to see and impelled to do works of grace on grace unto glory on glory because of the absolute orthodoxy, the impassioned zeal and the wholesouled sincerity of this great Pauline priest and orator.

CHAPTER XVI

THE WRITER

In view of all that we have seen of the tireless activity of the man, and the indefatigable industry that was manifested by Father Drum in the pursuit of literary, scientific, and sacred learning, we are prepared to inquire just what was the permanent achievement of all these labors. Speaking of material results only, what enduring use was made of all those remarkable gifts with which he was endowed? It is of course vain to measure a man's real worth to the world by the number of books he has written. None the less we are interested to know whether he has left some tangible monument of his industry.

It would, in fact, be scarcely credible that a scholar like Father Drum, after producing such a huge amount of writing for a score or more of years, should in the end have left actually very little in a complete and permanent form. It seemed, at the first investigation of his documents, that such was the case. He had spoken so little of his own writings, and at the moment, men had only seen him engaged in lecturing and preaching; while several of his controversies in magazines were attracting attention. The great preacher and lecturer had dropped from the ranks. Men did not think so much of the writer.

But as time went on it became more and more clear that Father Drum had all along been working with the same careful systematic order in his writing as he displayed in all his other activities, and that, had he been vouchsafed a few years more in which to develop his plans, there might have appeared at least a dozen useful volumes on various subjects, Scriptural, apologetic, and devotional, bearing his name and reputation.

As it was, however, nearly all the writing that he actually finished and published was in the form of a multitude of articles (the word is used advisedly) for magazines, and papers for encyclopedias. In all these he was following a plan which is more or less definitely visible now in his literary relicts. The articles were for the most part written with an eye to the time when they could be gathered together and retouched for publication as complete volumes.

At the end there were but three of these books in a more or less completed state. The first is entitled "The Shepherd Song," and is a devotional exposition of the Twenty-Second Psalm, so beloved in all his retreats. In fact, Father Drum had embodied in this booklet, which takes up but one hundred and sixty-two typewritten pages, many of the most beautiful meditations of his retreats, together with much of the Scriptural lore that he had woven into the explanations of the Psalms in the pages of the *Pilgrim*. It was meant to be a sacred keepsake of the happy days spent making the Spiritual Exercises.

The publication of this book was delayed only because Father Drum was looking around for a Catholic publisher who would bring out the volume in the specially attractive form that he wanted. This was something on the style of the publications of the Little Leather Library, and was suggested by a booklet sent him by a friend, entitled "Drum Taps," which were not taps of his own, but bits of verse from the war front. But up to the end his search was all in vain, although several offers were made him to have the book published in the ordinary form. It is probable too, that as time went on, the press of other work made him forget all about the "Shepherd Song."

The second volume that was almost ready for publication had the title of "John the Historian." The occasion and the purpose of his writing this book are explained in the introductory chapter:

> The wanderings of some of these Catholic writers [previously mentioned] are noted in the following pages, which the Encyclical *Spiritus Paraclitus* of Pope Benedict XV has inspired. . . .in order to forewarn and forearm the reader against shoals that may shipwreck one's faith. Many of these articles have already appeared in the *Homiletic and Pastoral Review;* and some of the contents of the book is drawn from our contributions to the *Ecclesiastical Review.*

Like many of his articles in these periodicals, the tone of the book is frankly controversial, and the entire volume is largely taken up with combating the errors that have found their way to condemnation by the Biblical Commission, or with refuting doctrines and interpretations of Sacred Scripture, that are not in harmony with the decrees of the Holy See. The book is full of sturdy defense of the traditional attitude of conservatism and loyalty to the least pronouncement of the authorities of the Church. "What then?" he asks, "Are we as free as they [the Fathers] in mystic exegesis? Quite so, if we presuppose the historical character of St. John's Gospel, and loyally follow the normative legislation of the Holy See." Many splendid tributes to loyalty occur in the following pages of the book. Again and again he gives vigorous expression of his condemnation of the too subservient attitude that some Catholic scholars sometimes take towards the writings of non-Catholic students of Holy Scripture. When all due credit is given to the wonderful research work that has been done in modern times by men who have the true scientific spirit, though lacking the true faith, we are not, therefore, called upon to concede to these writers the privilege of guiding us to the true sense of the Sacred Scriptures.

> Since we are forced again and again to make complaint at the contempt of Catholic literature on the part of Protestant science, we should more emphaticaly denounce such an attitude of disdain on the part of Catholic authors.
>
> This loyalty to heretics, and corresponding disloyalty to Catholic commentators, which one deprecates in *l'école large*, cannot fail of damaging the faith of the sheep of Christ. Why? For the simple reason, as Leo XIII says, in *Providentissimus Deus*, that the unerring meaning of the Sacred Books can be found nowhere outside the Church, nor can it be handed down by those who, being without true faith, fail to reach the kernel of Holy Writ, but only gnaw at its husk.

For the most part, the book is written in the style and temper of a vigorous debate, with the aim of presenting clear, cut-and-dried proofs, and with little care for the graces of diction or the flourishes of oratory. At times the very amenities of controversial style are forgotten, and it is war to the finish. One cannot help feeling that the author is perfectly right and convincing, but not a bit persuasive. However, were one to pass over these battle scenes, he would find pages of singular charm and beauty, in which is depicted the true mysticism of St. John's Gospel, and its value as a genuine history of Our Lord's life; there is a vivid and colorful exposition of the story of the Magi and their gifts; there are glowing tributes of loyalty to the decisions of the Church, as represented by the decrees of the Biblical Commission and other Roman Congregations; while his exposition of the idea of "life," "life eternal," in the Fourth Gospel, or sanctifying grace and the indwelling of Jesus in the soul, gives him the occasion for introducing a beautiful study of the sixth chapter of St. John, where is promised us the Food to nourish that Life Eternal in our souls. This section on the Holy Eucharist, composed in his very best vein of lofty spirituality and with deep devotional fervor, might well deserve to appear by itself for the spiritual good it may produce. The book was begun by Father Drum in October of 1920 and was finished in April of the following year. By September he was able to write to a friend: "I have a publisher for the book; but certain matters are yet in abeyance." Before these matters were finally settled, Our Lord had called him to his reward.

There is yet a third volume left in manuscript by Father Drum, not quite complete, which he had been preparing at about the same time that he was engaged in the writing of "John the Historian." This book, entitled "An Introduction to the Study of the New Testament," was intended to fill the need of a text-book of Bible knowledge in our college classes of Evidences of Religion. It was undertaken at the request of a fellow Jesuit, of the Missouri Province, who wished to prepare a book of choice selections from the Greek New Testament, and was anxious to have some one of authority in Biblical matters to write for him, by way of introduction, a concise scientific treatise that would be equivalent to a short handbook of the New Testament.

After the plan was agreed upon, Father Drum began about February of 1921 to work systematically at this treatise, and was engaged on the concluding chapter just before his last illness. He put a great deal of time and care upon this work, the more so as it was in line with his own long-standing conviction that there was a genuine need of such a text-book not only for our Catholic colleges, but for seminaries as well. At far back as September 1913, he had written in the *Ecclesiastical Review:* "We are excessively poor in Catholic treatises on New Testament questions of special introduction written in English."

As he was left free to determine for himself the limits of

this treatise, and as his co-worker had expressed the desire that the subject be developed with scientific thoroughness, without, however, making the book too technical for the grasp of college students, the "Introduction" finally evolved into a complete treatise of some twelve chapters, on the whole field of special introduction to the New Testament. The treatise is divided into two sections, in the first of which is drawn out with great thoroughness and an imposing army of documentary and patristic evidence, the whole argument for the historical worth of the Gospels, the proofs of the Divinity of Christ, and the relation of the Bible to the teaching body, the Church of Christ. The second section deals in detail with each of the Gospels and all the other books of the New Testament, analyzing all the great problems of the Biblical world with a thoroughness that really represents Father Drum's best efforts, and the ripest fruits of his learning.

It seems really unfortunate that he had not decided earlier to publish books like this, of positive apologetic value, and of real constructive usefulness. The great burden of his actual printed work was in the field of controversy, about which we shall speak presently. But in addition to the manuscripts mentioned above, many of Father Drum's conferences and retreat notes might form a volume that would be a welcome addition to our devotional literature. There is also material for a book of splendid First Friday meditations, in the monthly Conferences on the Sacred Heart, that were contributed for more than two years to the pages of The *Homiletic and Pastoral Review*. Besides, many urgent pleas have come to the present writer concerning the "Bible Talks" that appeared for five years in every issue of the *Pilgrim of Our Lady of Martyrs*, from 1913 to 1918, and in the *Queen's Work*, from 1918 to 1920. These "Bible Talks" were an attempt to revive an old Jesuit custom by which some portion of the Bible was expounded in a popular way on the afternoon of Sundays and holidays, although the name has now almost lost its original meaning. Father Drum took some of the Psalms and made a study of them for the *Pilgrim*. Many of these studies are of genuine beauty and full of rich devotion, and it is to be regretted that they did not receive a greater publicity than was possible with the circulation of the earlier issues of the *Pilgrim*. For the *Queen's Work* the studies were on the Book of Genesis, and proved to be too learned and technical for the pages of this periodical. Only nine articles were published.

Again, the hundreds of letters of spiritual direction that are easily available would fill several volumes, and might prove helpful and encouraging to many religious souls.

That it was Father Drum's actual intention to collect his magazine articles together and shape them into book form is clear from his own oft-repeated statement, and from the method with which he took care to preserve everything that he ever published. He kept several copies of each article as it appeared, and when a sufficient number had accumulated, had them bound together temporarily, intending to wait until he had leisure to retouch and rearrange the whole into a single volume. This was his actual method in the formation of "John the Historian" which in reality is but a recension of the "Scripture Studies" in the *Homiletic and Pastoral Review*, and of some of those in the *Ecclesiastical Review*. This was writing a book by piecemeal, as it were, watching its effect as the parts appeared and being prepared to modify or change whatever was necessary in the completed volume.

Up to his death, Father Drum had gathered together seven or eight bound volumes of his articles in the *Ecclesiastical* and *Homiletic*, and also a bound volume of scattered articles that appeared in various other magazines and which he had some reason for preserving. One might imagine that these Biblical studies in the monthly issues of the *Ecclesiastical* for ten years, from 1910 to 1920, and in the *Homiletic*, from 1919 to 1922, would be sufficient labor to occupy the time of a man who was otherwise kept quite busy, as we have seen, with professorial duties and the work of preaching and giving retreats. But in addition, during all this time, he wrote contributions to numerous other magazines, and, in fact, these scattered articles represent far more than the total output of the average student who takes an interest in magazine writing. The wonder is how he found the time to do them all. There were from time to time contributions to *America*, in later years mostly book-reviews, though even a book-review means a considerable expenditure of time to one who was as painstaking as he; there were articles in the *Catholic World*, *American Catholic Quarterly*, *Expository*

Times, Truth, Queen's Work, Bibliotheca Sacra, Messenger, Messenger of the Sacred Heart, Newman Quarterly, Notre Dame Quarterly, and in various college reviews.

Yet, incredible as it may seem, this was by no means the total output of his busy pen. There was far heavier work going on most of this time. From 1910 to 1912 inclusive, Father Drum produced the articles under forty-four headings in the "Catholic Encyclopedia." Some of these, it is true, are very short—only a paragraph or so—but others are of considerable length and represent an untold amount of labor in the preparation. The article on the "Incarnation" takes up twenty-one columns in the close type of the first edition of the Encyclopedia; that on "Manuscripts of the Bible," twelve columns; that on "Psalms," twenty columns. His name does not appear in the first six volumes, all of which appeared before and during the year 1909, the year following his return from abroad. Besides his own articles Father Drum was also asked to edit several by other contributors. Later, in 1918 and the following year, he was engaged to write thirteen special articles for the "New Encyclopedia Americana." In this new edition, his article on "Christology" is thirteen columns long, that on "Manuscripts of the Bible" takes up more than fifteen columns, while about ten columns are given to each of the articles under "Renaissance," and "Resurrection." The articles on "Immaculate Conception," "Index Librorum Prohibitorum," "Jansenism," "Religion," "Trent," are also in the list of those for which he is responsible.

Perhaps it is true to say that as a writer Father Drum was best known for his monthly contributions to the pages of the *Ecclesiastical Review.* The most diverse opinions have been passed on the character and usefulness of his work as editor of the department of "Recent Bible Study" in the review. That his studies were scholarly and accurate no one denied. They were for the most part written in a free style, almost like letters. They show how anxious he was to keep up with all the current literature of the Bible. His researches ranged over all Europe and America, since his ready command of the German, French, Italian, and Spanish languages opened to his reading and critical study all the publications that had to do with Sacred Scripture, with theology, or with archeology. His knowledge of Latin,

Greek, and Hellenistic, and of the Oriental languages gave him a special authority in this field.

But it was almost inevitable that one of his militant character and temperament was not long to rest satisfied with merely reviewing the books and writings of other scholars. His studies very soon began to take on the form of an outspoken expression of his own opinions on the subject under review, and of a criticism that was at times caustic and severe. In one of the issues for 1912 a rather warm discussion arose over the relatively unimportant question of the interpretation of St. John ii, 4, "Woman, what is it to me and to thee?" and was continued at intervals for several numbers. Soon afterwards a still warmer controversy arose over the question of the *Parousia*, or Second Coming of Christ, in connection with some criticism of his own interpretation of 1 Thessal. iv, 15-17, in the article under "Thessalonians," in the "Catholic Encyclopedia."

For the next two or three years Father Drum continued to fill the department of "Recent Bible Studies" with more constructive and useful work. He was also requested by Father Cuthbert Lattey, S. J., one of the editors of the "Westminster Version of the Bible," which is still in course of publication, to undertake the version of the Gospel of St. John. But after a prolonged correspondence it was found impossible to agree on a concerted plan, and Father Drum's version was never completed.

About the time that Father Drum was delivering the third course of lectures before the Brooklyn Academy, he also began, in the pages of the *Ecclesiastical Review*, to berate the Professors of Harvard, Brown, Chicago, and other universities for the same un-Christian teachings that he was exposing in the Brooklyn lectures. The editor pleaded in vain that such attacks were not likely to be productive of much good. They would not be read by those against whom they were directed. The *Review* was intended for circulation among the Catholic clergy exclusively, and the benefit that might accrue from warning Catholic priests against the vagaries of non-Catholic professors was disproportionate to the effort involved. The following letter was written by the editor about the beginning of this campaign:

8 Jan., 1917.

Dear Father Drum,—

. . . . You asked in your last note whether I thought

you kept overmuch to Christology in the "Bible Study." No. I deem that none is a better judge than you as to what particular phases of Scriptural studies will serve to animate our readers with reverence and love of the written Divine Wisdom as suggested by the current new and important literature published on the subject. I believe we all admire your critical acumen, your thorough knowledge and radius of expression in dealing with up to date matters in your special field of study. If you want me to find fault, or rather if you want me to find some point in which idealists like myself would seek additional perfection in the Department of Biblical Study, I would ask for emphasis of the eirenic rather than of the polemic elements in your habitual strictures. The gentle courtesy that corrects a fault without caustic penalty has always seemed to me a specially attractive dynamic, though I keenly relish also the occasional (good-humored) stroke and touch of the sharp whip that rouses the blooded runner. But you are guided by prayer and your work is for God's honor so He Himself will direct your ways in your chosen exalted field.

Wishing you the joys and blessings of the New Year, I am, dear Father Drum,

Faithfully in J. C.

H. J. HEUSER.

When, however, Father Drum finally decided once more to go on the war-path against another Engish Catholic writer, in a matter that promised to grow into a combat of serious proportions, the editor of the *Review* took genuine alarm, and could not be induced under any consideration to continue the publication of further controversial articles. After a long correspondence on the subject, Father Drum finally withdrew from the *Review*. Shortly afterwards he was invited by the editor of the *Homiletic and Pastoral Review* to take up the Biblical Department in that periodical, in additon to continuing the "First Friday Conferences" that he was already contributing.

But that the difference of opinion between himself and his opponents was not permitted to become a personal one, and that he was understood aright even by those who disagreed with him, is clear from this excerpt from another letter of the distinguished editor of the *Ecclesiastical Review:*

Dear Father Drum! I had no idea he was ill or suffering until I learned of his death. Yes, he was a scholarly and hardworking religious, and I had reason to know that he

> was also at heart of a very humble and lovable nature. Somehow the military spirit inherited from his father made him incline to polemics in a way that was contrary to the eirenic aims of this magazine. I tried to explain to him and his brethren; but his love of truth was greater than his appreciation of peace, and our judgments as to opportuneness of attack and defense differed. May his dear soul rest in peace. I frequently think of him at the Altar because I may have given him needless pain by my words.

On another occasion the same genial editor bespoke his long continued personal admiration for Father Drum.

> I usually take people as they are and look to their sincerity and their aims, or lack of them, and rate each one accordingly. Their mode of action or their external mannerisms affect me not at all. That is why I revered Father Drum for his loyalty to truth, for his religious life and character, for his real humility of heart, and forget his spirit of combativeness and the brusqueness of style in his writings. Some people criticize his style, and say he was artificial and pompous in manner. That is no criticism at all. One might just as well criticize a man for being too tall or too short; his manner and his style are just as much from nature. The style is the man. But sincerity and zeal for truth and love of hard work were unmistakably his best qualities—and besides, I often wondered, as I dealt with him in personal conversation, or heard the praises of others who had come in contact with him, how it was even possible for a man of such a really lovable nature, and of such genial social qualities to write in such a vein of invective when in controversy.

Although the controversial articles of Father Drum, in the *Ecclesiastical Review,* were followed with keen interest by many of the clergy, great numbers of whom used to write to assure him of their enthusiastic support and to applaud the manliness and vigor of his defense of conservative Catholic opinion, it is yet true to say that this phase of his activity was considerably misunderstood in some quarters, and he was often subjected to severe criticism. This of course was inevitable in any event. People said that his method of attack was of the old style, and not in harmony with the easy affability of modern controversial courtesy. His was the polemics of Jerome and of Chrysostom against the heretics of their day, or even against those who showed the least trace of falsehood or of disloyalty to the teachings of the Church. It was a polemics of merciless war against error in any form.

Since Father Drum's death, a high authority in the teaching of Sacred Scripture writes from Italy:

> I do not care to write an appreciation—I did not know Father Drum personally. This however was clear, that his spirit was an apostolic spirit; and he would with St. John have run out of the house on seeing Cerinthus, lest the structure might by some chance fall in and he be crushed with the heretic. He would have met Marcion too as did Polycarp: "I know the first-born of Satan." Indeed those who were hurt by him never realized how much they had provoked him, and in points of dogma and scripture he was right. There are some books in circulation. . . .in which a novice in theology may see heresy stalking. God rest him! He seemed to take to himself St. Paul's advice to Titus (iii, 10): "A man that is a heretic, after the first and second admonition, avoid: knowing that he, that is such an one, is subverted, and sinneth, being condemned by his own judgment." He seemed to have caught some of the spirit of the great Apostle to the Gentiles.

From Spain comes a similar appreciation, from a Professor of Sacred Scripture. The letter is written in Latin, and part of it is translated as follows:

> It seems to me to be a most welcome plan of your Superiors, to prepare a biography of the lamented Father Drum; and I regret that I cannot take a fuller share in the labor. I had some correspondence with Father Drum, and wish now that I had preserved it. From these letters, from all that I heard from others, and also from Father Drum's own writings, I easily gathered that here was a splendid scholar, well-versed in Sacred Scripture, a man of depth and wide erudition. I was in Rome at the Biblical Institute when Father Drum died; and this I can affirm, that the news of his death was a great sorrow to all the Fathers there.

At least let him be conceded the tribute of admiration for his absolute fearlessness in defense of the truth. If he erred at all, it was in being too conservative in his opinions, and too ready to march to the attack where another might have bided his time. Indeed, he showed little inclination to soften the force of his blows, or tone down the rigor of his criticism. He felt that he was dealing with opponents who were themselves responsible men, and that the objective truth to be defended was of far more importance than the personal feelings of those who might be the cause of damage to the faith of the people. When once aroused, he could be simply relentless. For the Pontiff's ringing appeal, the old war-cry of Pope Pius X, appears to have taken full possession of his mind and heart: "Remember, stand by the traditional doctrines of the Church!" Not one who ever came in contact with Father Drum, even of his opponents in controversy, but bears witness to his sterling Catholicity, and, in his particular field, loyalty to the Church resolved itself into an unswerving adherence to every decision of the Biblical Commission.

Often and often he expresed his amazement at the complacency with which some Catholic writers ignored more or less completely the publication of such decrees, as if they were not meant to be the authoritative norm of opinion on questions of dispute in the Sacred Scriptures. He urged upon all priests the duty of following the proceedings of this and other Roman Congregations, as published in the official organ of the *Acta Apostolicae Sedis*. He himself took the trouble to copy out for his own use, and kept the full list of decisions of the Biblical Commission from June 19, 1911 to April 23, 1920. He kept abreast of developments of Biblical and theological studies in Europe and America, and was ready instantly to point out and attack any deviation from that high standard of loyal orthodoxy which he always held up for himself and others. He attacked it never gently, his opponents will have it, never tactfully,—yet in every discussion he left not the least doubt as to the real issue that he had raised, or the sincerity and conviction and breadth of information with which he supported his own position. Though the warfare created inevitable irritation in many quarters, yet one definite achievement, valuable to the Church, must be noted in any estimate of Father Drum's life work. As a result of his constant watchful attitude, it may be said that the fear of God descended on many a book-reviewer and writer in this country and abroad. They became more cautious in hailing the advent of any new discovery in the field of Biblical investigation, and more careful to examine the claims of pseudo-scientific critics whose whole aim in life seems to be to find new ways of dividing up the Bible and explaining away every element of the supernatural or the miraculous in the Holy Scriptures. This may be stressing a point rather strongly, but it is true to say that an attitude of greater deference to the decisions of the Holy See was created by his efforts, and consequently the spirit of loyalty was strengthened in many who were being led astray. This much was surely gain.

CHAPTER XVII

RETREATS

The giving of the Spiritual Exercises of St. Ignatius, or retreats, as they are usually called, is one of the chief works of the Society of Jesus. There is no need to dwell here on the potency of this wondrous instrument of spiritual warfare, or on the Church's official attitude towards the Exercises as a means of sanctification and of perfection. They are here alluded to, merely to call attention to the fact that a man who was devotedly loyal to every work that was proper to the Society would surely be expected to take a keen interest in, and show unbounded enthusiasm for the work of retreats. Such is emphatically what we find in the case of Father Drum. With his usual systematic precision and orderliness, he kept a record of every occasion when he conducted the Spiritual Exercises. The list begins with the years 1890-92, the time of his Noviceship, when he conducted the Exercises four times. The next entry is for the year of Tertianship, when he gave six retreats and missions all told. The Scholastics during the course of studies are not usually called upon to conduct any retreats; nor had he any opportunities for work of this kind during his two years abroad, as he mentions regretfully in many of his letters. But in 1908, the year of his return to the United States, he conducted nine retreats altogether, to the Sisters of Charity at Mount St. Vincent, N. Y., to the Ursulines at Bedford Park, N. Y., to the Sisters of Mercy at Providence, and to other Religious communities. In the following year, 1909-10, he conducted the Spiritual Exercises sixteen times. But the year 1911 is marked with sad but expressive brevity by the single word: "Nerves," to indicate the breakdown which made him give up all work for that summer. With this one exception, the retreats of every year are listed singly, month by month, from May 1910 to September 1921, making in all a total of one hundred and twenty-two retreats, and an average of more than twelve retreats a year.

No further comment is necessary on the zeal and self-sacrifice that this work involved. It must be remembered, too, that his other duties gave him ample reason, had he wished it, to decline the offer of work of this kind, and that therefore the giving of retreats was for him purely a labor of zealous supererogation. But there was none dearer to his heart. "How I long to be able to get back to this work," he wrote from Innsbruck.

Furthermore, it must be remarked that these retreats, contrary to a general impression, were not always made by the same class of exercitants. There was a considerable variety in the character and make-up of the groups to whom Father Drum gave the Spiritual Exercises. Besides the number every summer to Religious Sisterhoods, and the many occasions when he conducted retreats for school-teachers, business women, and students, there were the Knights of Columbus retreats to daily increasing crowds of men in the Cathedral of Wilmington, Delaware, in May 1919; in the Cathedral of Richmond, Virginia, in June, 1921; and also in Ithaca, Providence, Rochester and Syracuse. In August, 1916, there was a retreat conducted by him for the Alumni and the Law and Medical Students of Georgetown University. There was one retreat to the candidates for Ordination at Overbrook, Pa., in May 1916; one to the priests of the diocese of Nova Scotia in 1918; two to the priests of the Fall River diocese, at Holy Cross College, Worcester, in 1920; and two to the priests of the Hartford diocese, at South Norwalk, Conn., in 1921.

A priest who was among the exercitants at the Fall River clergy retreat at Holy Cross speaks of the enthusiasm and the success that attended Father Drum's splendid efforts at these retreats. Among other glowing eulogies, he writes these words of appreciation:

> Worn and weary with the heat and burdens of the year, we went to that annual retreat with the serious purpose of entering into ourselves to see if so be we might know how far we had wandered from our true ideal the great High Priest, Jesus Christ Himself, whose Priesthood we enjoy. And there in pursuance of that exacting, soul-stirring duty, we became acquainted, many of us for the first time, with the lovable Father Drum.
>
> Would that the gift were mine to portray him as we saw him and knew him and learned to love him during those happy retreat days. How shall I describe him? He was a

man of winning personality, of cultured mind and of intense spirituality. Rich possessions in any man, but rendered richer still in him because of his generous willingness and marked ability to use these gifts for the delectation and profit of his fellowmen. A ready flow of beautiful language together with a well-trained, modulated voice made it seemingly easy for him to impart, and easy for us to receive the fruits of his well-stored mind. Add to all this a never-failing, sympathetic considerateness for his audience, and you can understand how he soon so won our love and confidence, that although five times a day he spoke to us in conference, he left us each time longing for more.

In Sacred Scripture he was at his best. *Facile princeps*, he spoke as one who knew. The Old and New Testament and the history of the Church were to him as an open book. The certainty and clearness and conviction with which he expounded the truths of our religion engendered in our minds and hearts a deeper appreciation of the reality of the Faith that is ours.

Another priest, of the Philadelphia archdiocese, also writes a few of his own recollections of the Ordination retreat at Overbrook, in May of 1916:

June 29, 1922.

Dear Father:—

Your beautiful tribute to the late Father Drum in the May number of the *Sacred Heart Messenger* revived in the writer's heart the spiritual fervor of the days immediately preceding his ordination. Those days were indeed days of inspiration, overflowing with the spirit of the Apostle to the Gentiles and blessed with the zeal of Chrysostom and Jerome—the holy hours of Father Drum's retreat!

His spirit was indeed the spirit of an *alter Christus* and "another Paul." From him my love for St. Paul was truly intensified and his vivid expositions of the Pauline Epistles created in my soul a vehement desire to search more enthusiastically for the unlimited treasures contained in those letters.

You have indeed read the great soul of him whom I reverenced as a true priest, an eloquent orator and a writer whose pen was wielded nobly for his Church as his father's sword was unsheathed for his country. May his soul rest in peace and may the inspirations which I received from his retreat *never cease* to make me a true follower of Christ and an ardent admirer of St. Paul.

Wishing you the best of health and happiness and again expressing my worthy esteem for your kind tribute to our mutual exemplar of the Catholic Priesthood . . .

However, it is probably true that Father Drum's retreat work is best known for his long connection with the retreats for teachers and for business women, conducted yearly by the Religious of the Cenacle in New York. One of the Sisters very kindly furnished the following facts in regard to these retreats: For a period of ten successive years, from 1909 to 1919, Father Drum conducted in all about fourteen retreats at the Cenacle of St. Regis, New York. His first retreat at the Cenacle, in 1909, it was understood then, was also the first retreat for women that Father Drum ever gave. It was a year after his return from his sojourn and year of study in the Holy Land. The house journal of the Convent for September 4, 1909, says of this first retreat:

> From the opening of the retreat last evening, Father Drum held the retreatants spell-bound as much by his evident piety and earnestness as by his words so eloquent and full of life. We are rejoicing that so many souls are here to profit by the exercises. Eighty retreatants were at the Cenacle at this time, and it is remembered that the spirit of enthusiastic appreciation and gratitude reigned in the hearts of all at the close of the retreat. Before leaving Father Drum gave an inspiring Spiritual Conference to the members of the Community on "The Disciples of Emmaus."

In 1912 Father Drum gave two general retreats at the Cenacle of Newport, R. I., and the following two years his retreats at the Cenacle of New York were especially for teachers. The general retreat over Labor Day, 1915, was so overcrowded, that his time and strength were very much taxed, and since he was to conduct the Labor Day retreats of following years, it was thought well to have two retreats instead of one at this particular time of the year. In 1916 and 1917 Father Drum preached two retreats in succession in order to accommodate the hundreds who flocked to the Cenacle to follow the Exercises under his direction. It was at this time, during these years, that he gave what he called his "Pauline retreats"—wherein, in a luminous way, he illustrated the Exercises of St. Ignatius by applications and quotations from St. Paul—describing St. Paul "in the ring"—fighting the fight—St. Paul "running the race" (Phil. ii. 16)—"The Rules of the Race," 2 Tim. ii. 5—"The Call of the Mystical Christ," etc., etc.

"Come to the retreat and get some Paulinity," was the wording of one of his invitations to make the Exercises.

Father Drum asked in a striking Conference on John the Baptist given to the Community of the Cenacle in 1916:

> What is our life? *To make right His way.* Our life is true in so far as it is right in regard to Christ and to Christ's way: To teach the people to reason rightly, by the mercy of God, unto the remission of sins—to bring the life of Christ to those that sit in darkness and in the shadow of death.

This it might be said was the keynote to Father Drum's retreat work. He labored to bring Christ to souls—to prepare His way. His method of doing this was colored and characterized, of course, by his own personality.

> Quite vividly I recall [writes another correspondent], the very first conference of my first Retreat under Father Drum, a Retreat for Teachers at the Cenacle, New York, wherein he pictured in a startling way what proved to be the identical circumstances of his own death. I recall also, that for the first day or two I thought him a "cold Intellectual," and appreciated that he made his appeal to the highest. However, when we came to the Holy Hour, he gave vent to the most sublime emotions and the tender pleadings of a consuming zeal,—and then I got another impression which later days and years served only to confirm. As in all reverence I write of him today, his saintly likeness looks down upon me and I can see the words he had written in his own hand beneath, so beautiful and consoling: "If God be on our side, who dares be against us?"
>
> Holy Hour was the time in which his physical body seemed to droop under the strain of his overpowering emotion [writes a lady who had made the retreats under his direction for several years in succession]. I remember on one occasion when he was speaking of the Agony in the Garden, how his forehead became beaded with large drops of perspiration, his face became flushed and finally the tears poured from his eyes, although his voice scarcely betrayed the anguish he was enduring.

Another has this recollection of one of these retreats:

> Father Drum's personality was truly magnetic. He reached each and every one of us, knew us all. There were a hundred making the Retreat. I was absent for a day. He missed me, and immediately inquired. He knew a struggle I was having that made me ill. Such

> fatherly love,—It's too beautiful to put into mere words. A heart like the dear Lord's own, full of love and tenderness.
>
> Bring out at some length [one asked of his biographer] what was always regarded as a most striking characteristic of Father Drum—a fatherly interest and tenderness for souls of those who looked to him for guidance. It is unlikely that those who knew him as preacher, lecturer, writer, knew him in this capacity as well.
>
> In many letters, mention is made of his support in sorrow. The marvel is at the burden of grief he bore for so many of us.
>
> Truly he is a man of God [writes the Dean of a well-known college for women]. Father Drum's campaign may be summed up: "He came, they heard, he conquered!" The students beg me to engage him for the next year. He has an inimitable way of dealing with them and of gaining them. By frequent Communion and in numerous other ways he worked untold good. His original and effective use of the Bible, his interpretations and very practical applications were also remarked.

About the beginning of 1918 Father Drum decided to extend the range of his activities in the matter of conducting these retreats, and accepted the repeated invitations of the Religious of the Sacred Heart of Mary, of Marymount College, at Tarrytown-on-Hudson, to conduct the Spiritual Exercises for their benefit, and also for the large group of alumnæ and business women who were annually attracted to this beautiful spot for a spiritual vacation. The attendance at Marymount was even larger and more enthusiastic than ever before. Among the few letters that were found in Father Drum's room at the time of his death—he very seldom kept letters after they were answered, it would seem,—was one from a lady who had made the last but one of these retreats at Marymount, that of August 27, 1921. It was not written to him, but to a mutual friend, who thought it so interesting an account of the workings of grace on a somewhat unwilling soul, and the effects of the retreats and the nature of Father Drum's influence on those with whom he came in contact, that it was forwarded to him for approval. We do not know if he ever read the letter—it must have reached Woodstock shortly before his last illness—but with the courteous permission of both parties concerned, parts of it are here appended:

. . . . At last I am here with news from the "front" which I know you are anxiously awaiting . . . I wrote to Father Drum, telling him I had decided to make the retreat, but received no answer. This, of course, set my wild imagination to doing all manner of calisthenics, and I quickly arrived at the conclusion that the good man was aweary, and bored to death with me and my woes. Everyone has his own dreary road to climb, and if, out of the goodness of his heart, a companion pauses by the wayside to give the helping hand to one in distress, it doesn't follow that that one should drape herself about his neck and "holler" for a hitch to the stars.

And so . . . being very heart-sick and sore, I said to myself, "I'll go home to my own little cot and fight it out alone; it's *my* battle; why should another bear the burden . . . A hundred times I said, "I'll go home," but Saturday, August 27, 9:45 a.m., found me with a few faithful satellites at the N . . . station, awaiting the coming accommodation train for New York. I didn't know exactly how to reach "Marymount." I was tired to death, because of the grand finale the day before at N . . . ; all I knew was that I was to get off at West Point and take the ferry over to Garrison where I was to get a train for Tarrytown.

At Garrison, on the opposite shore of the Hudson, the train was not due for forty minutes. It was a pretty little country station, but, as Father Drum says, there was nothing to do but "set and think," and as I didn't dare think, I "jest set."

The day was beautifully clear and sunny, but decidedly humid and hot. After the clear, cool, bracing air of the mountains, my body began to wilt with my spirits, and when the train pulled in, crowded to the gunwales with returning children of Israel, my cup of bitterness began to run over, and I longed for the freedom and seclusion of home, sweet home. . . . At Tarrytown, I secured a taxi, and soon arrived at "Marymount" the Beautiful. Extensive, well-kept grounds; grass green velvet; old, old trees that probably witnessed many a bloody battle between the red man and the conquering, ruthless white;—beds and beds of flaming geraniums and scarlet salvia (in honor of the Immaculate Heart)—this is "Marymount." Along a "gravelly" path we rolled to a side entrance. I stepped in through hospitably open doors, into a wide, cool hall. Peace and the quietness of sanctity permeated the place, and surrounded as with a holy light, the sweet, placid-faced nun that glided down a broad staircase to bid me welcome.

The chamber assigned me was a "cuddley" bit o' white room, with a white enameled iron bed, white bureau with a tiny mirror in which one saw only part of one's countenance, and a white wardrobe backed up against the wall at the foot of the bed, apologizing for being there. I sat down on the bed dejectedly, dropped my bag just where I sat, and thought, "Well, here I am at last. What next?" I was hungry and tired; my eyes began to fill and a big lump arose in my throat, but I bravely swallowed it, hoping thus to appease my appetite for food. Finally, I arose, opened the door, and ventured down the corridor. I found a very pleasant girl who had made the retreat the year previous under the direction of Father Drum, and who promised me a wondrous "treat," all of which I listened to in polite patience and much skepticism, with a raging spirit within. . . .

In one, twos and groups, the "girls" began to arrive. I watched them from my window and noted the different styles of architecture (again W. D., S. J.,); high, low broad, shallow, old-style, new-style, sweet faces, stern faces, all serious. The halls and corridors began to resound to hurrying feet and laughing voices of friends renewing old friendships. Outside my door, the atmosphere became heavy and humid with a shower of Italian "a's," exclamation points, and startling adjectives, interspersed with much "lawfter."

After dinner, we had recreation until eight o'clock, when we had meditation in the chapel, instructions as to our behaviour while on retreat, confessions, bed. I was anxious to catch a glimpse of the "Fawther," and entered the sweet little chapel, tired, heart-sick, and home-sick, curious and cynical. Such an awful attitude to commence a retreat!

Then, behold, he strode up the aisle, and on the altar, under the soft glow of candle and electricity, I saw the man of men. He was not at all what my mind had drawn. You showed me that little snapshot, and from it I had builded something quite different. He spoke. His voice is beautiful in quality, forceful in tone of denunciation, sweet and tender in sympathy for poor suffering humanity —a great-hearted friend and leader; a fighter to the last ditch; the leader of a forlorn cause, with patience without limit. . . .

That evening, after along, tedious wait, I went to confession. He awaited us in a room off the chapel corridor, where, before a little screen, one spilled out all the messy faults, crimes, and misdemeanors of a vacation spent not altogether wisely but too well. I wanted to go to confession before I met him, and too, I wanted to receive Holy Communion each day while I was there. I wager you. . . .dollars to doughnuts, I had hardly uttered three sentences when he knew me. No, of course not; he didn't advertise the fact, and he told me he first discovered me at conference, because I was the only one who wouldn't look at him and smile . . . The confession was not very satisfactory. I wanted to make my escape, and he couldn't very well keep and question me without letting me know that my identity was known to him. So he blessed me and let me go. . . . Up early next morning (Sunday). Mass and morning prayers from eight until nine, then breakfast, freedom to tidy one's room until ten; meditation with Father Drum, adoration (each one had her special half-hour), conference at twelve, dinner at one, adoration, rosary at three, instructions at four, Benediction at five, dinner at six, recreation until eight, instruction, and night prayers, confessions, bed any time ready.

Father Drum sent for me Monday a. m. but I was strolling about the grounds, and they [the nuns] failed to locate me. I went up at noon with ———, received an introduction, though he did not wait to get it, but shook hands. He had to interview several retreatants so I waited until evening recreation, and had a whole hour of his precious time. The weather had become very hot, and my heart was too full of pity for the poor man. I wanted to get away. I was afraid of him. I can't tell you why. I don't believe I could ever be very friendly with him. He is too far away—above me intellectually, as the heavens are above the earth; an intellectual giant, with the languages of the world on his tongue's tip. What can such an "earthly worming". . . .have in common with this "Master Mind!" I tried to tell him something of the past.

Gradually, all the cold, cynical thoughts disappeared, and the last morning, at the final words, all my bad thoughts, like the mist before the morning sun, arose and melted away. . . . I left Father Drum a small token of my appreciation and esteem—I only wish it could have been larger—bade him good-bye—he gave me his blessing, and we left a little after one, in Miss N.'s auto. We had a beautiful ride home along the majestic Hudson and silvery lakes.

The priests of the diocese are enthusiastic over Father Drum and his retreat. One of the nuns says it was the best retreat she has made in thirty years. But I can say much, much more; it's the *best I ever* made.

> He had me tied, thrown, and branded (W. D., S. J.). I got everything I asked for on retreat. . . . When I arrived home, I found Father Drum's letter awaiting me. It was in answer to mine, telling him of my going to "Marymount." He thought I had gone home. It was a sweet letter of encouragement, and so cleared the atmosphere of much electricity.

Similar expressions of grateful recollection occur in many letters. A nun writes:

> I heard Father Drum say once, in the earlier years, after his return from abroad, that he had only one retreat and always gave the same one; but I think no written notes can give an idea of the vividness of life and color, and sincere spirituality that he threw into his speaking. . . . I may truthfully say that for me it was and will probably always remain *the Retreat.*
>
> I made Father Drum's Retreat [another nun writes] ten years ago. It is still more vividly in my mind than any I ever made since. We never expect to have another like Father Drum.

However, nothing annoyed him more than the impression which some people took away, that his setting and the original presentation of the great truths were a departure from the Ignatian ideal. He repeatedly insisted that he studied to follow the Exercises very closely.

A religious of long experience and keen judgment sums up the character and influence of Father Drum in a brief but remarkably clear analysis:

> With the nuns, Father Drum's powerful influence seemed to flow from his Christly manner, his keen insight into their religious life with its joys and sacrifices, and his fatherly and sympathetic interest. His words were direct, frank, convincing and practical. He urged all to a close imitation of Christ, even to the extent of suffering, humiliation and failure, if these should not interfere with God's glory. His exercitants felt the inspiring influence of one who was filled with a personal love of Christ.
>
> I have often heard Father Drum criticized for being too learned in his talks [writes another Religious]. Yet no one was more direct and to the point than he. If he used unusual words, he always made their meaning clear by paraphrase. He was never merely emotional. Yet he could rouse one to enthusiasm and ardor by his startling way of presenting things, his persuasive tone of voice, his glance,

> and the charm of his manner. Really, when one's fervor cooled, there remained the convincing force of his arguments, the reasoning clearly set forth, rockbottom facts, the meat of doctrine,—and it was this, I think, which effected so much good.

In these retreats to Religious, Father Drum was scrupulous to give of his very best efforts. It was repeatedly remarked that he never spared himself, and perhaps it was this genuine willingness to be of help that gave him his greatest power over others. He once spoke of a complaint that a certain Religious Superior had made about some director who had been unwilling to hear confessions except on one day and at a stated time during the retreat. That made him very indignant. "He should have been reported to the Provincial," was the reply. His own attitude was that the director who is giving the retreat is there to spend himself and to be of service to all at all times.

Father Drum's influence as confessor can, of course, never be measured by any human standard, but we may understand something of his method of dealing with souls and the ideal he aimed at from these words of one of his conferences:

> Confession was never instituted as a mere tribunal wherein the soul is nothing more than accuser and accused. No, Confession was also meant by Jesus to meet the greatest human need we have, the need of a confidant who is sure to keep our secret; the need to tell our weaknesses to one who will not make us suffer for them; the need of our dearest of friends who does not come too near; the need of one who will give us an honest judgment without prejudice for or against us and without the possiblity of gain for himself.
>
> There must be firmness [he writes in another connection, but pertinently to the present subject], but firmness must be tempered by love, and of love, St. Paul writes, "All things she excuses, all things believes, all things she hopes, all things endures" (1 Cor. xiii, 7). Note the gradation. The evidence is at times so clear that there is no *excusing*. Then love *believes* that it must be well. No, the evidence is too strong for such *belief*. Then love *hopes* on for the best. Not in this case; it is *hopeless!* Then love *endures* it! "Love never fails!"

There is little room for doubt that his zeal for this apostolic work of retreats, sprang from no other motive than from that high reverence which he felt for the Religious vocation, and for the glory of a life of consecration to God. Says one Religious:

> His talks were full of beautiful expositions of the trials and triumphs of religious life, the spiritual glory of the mystic espousals, and the reward that even in this life is granted to those who persevere in the love of the Crucified Bridegroom.

The conferences especially reveal an extraordinary gift for imparting thorough, necessary instruction on the Vows of Religious life, on Confession and Communion, on the government and rules of various Religious Orders and Congregations, on the duties and privileges of Religious and of priests. One Religious wrote at the end of some notes on a particular conference, in a book that was in all likelihood not intended for any other eye: "What a clear, well-defined Conference. Thank God!"

Another Religious thus sums up the results of a retreat:

> The lessons I learned from that first retreat, driven home ever more forcibly in each subsequent one, were first and foremost, that Jahweh is my loving Father, that I must ever trust in His sweet Providence, that I must learn to think things out calmly,—be guided by reason grace-led, and will grace-driven, and never by my grovelling emotions; that Our Lord is worth my while to follow, and His call must get a generous response.
>
> Other things, too, have never been effaced,—the proofs of the Real Presence, a sermon on mixed marriages, the beauty of sanctifying race, how grace builds on nature and is not destructive thereof, the greatness of the heart of Paul.

There are literally scores of appreciations that have been written since his death, that leave no doubt as to the unusually vivid nature of the impressions that were produced by Father Drum's retreats, the deep spirituality of his influence over souls, and his consummate mastery in the use of the Spiritual Exercises of St. Ignatius. To quote from various letters:

> They [the Religious] are charmed with the Retreats, it is all based on the word of God, solid and instructive.
>
> We were delighted to see he followed so closely the Ignatian method and ideal; some of your Fathers have disappointed us because they tried to be modern. What is the greatest grace we have to be thankful for this year? Father Drum's Retreat.

> A clever dealer in the unusual in thought and in phrase, he has a way of expressing the truths of the Exercises that recurs to the mind even long after the retreat.
>
> I especially liked the frequent recourse to Holy Scripture, especially the Old Testament, and the reality he could impart to the composition of place by describing the actual scenes, which he had himself visited; and he possessed to perfection the gift of mixing grave and gay in just the right proportion.

More of this last characteristic later. These are but a few random selections from scores of letters that poured in soon after the announcement of a possible biography of Father Drum. It was quite natural also that many of those who made the retreats would be at once attracted and won by his striking presence, his vivid personal manner, the stories of his own boyhood and parents and teachers, and the narratives of his travels and personal experiences. These, however, were but the external adornments of an enthusiasm for souls and for the cause of Christ that was too evident to be missed.

> In beginning this Retreat [he would say], let there be something solid to build upon; let there be a firm hope and a firm love—not fear. That hope is founded upon the principles which we are going to talk about during this retreat. The first thing is, therefore, great-mindedness, not little-mindedness, great-heartedness and magnanimity, not little-soulness and pusillanimity. You can be excused if you have a little body, but you cannot be excused if you have a little soul. You ought to have a great big S O U L —So, let no fear interfere during this retreat. Don't be afraid of me. I am armed with nothing more dangerous than an umbrella. Don't be afraid of yourself. I should be afraid of being afraid of self. Therefore, first, no fear, but big-heartedness; secondly, pray during the retreat. Pray for all; pray for each commencing the retreat; pray for me. Pray that I may say the right thing and say it in the right way. The right thing to say is the Christly thing to say to each of you—to say the Christly thing in a Christly way—in the way that will help you most, not in a way that will rock you to sleep. . . .nor sandpaper you, but rather, help you to love our Lord, because the mission of the priest is a Christly one. The priest should be another Christ in each mission and Retreat.

It may not be out of place to quote a few of the stories, taken at random, to illustrate his method:

> Some years ago, while crossing the Gothard Pass between Switzerland and Italy, I marked the tugging and the toiling of the great engine as it drew the mighty train up the steep incline. The road was a rack and pinion road. As the engine tugged its way up the ascent it pinioned the train against rung after rung of the rack; nor did the train ever slip. In the end the engine scudded away with an even motion. It seemed to me I could hear that mighty engine talk. It seemed to say: "I *think* I can; I think *I* can; I think I *can.*" And then, up on the level: "I knew I could." You will often have to say "I think I can." The grace-power of Christ will always be there to draw you up against the dragging power of the world. The time will come, sooner or later, when you will look down at the past and wondering at the obstacles gone by you will say with joy and security and enthusiastic love of Christ: "I knew I could."

The following excerpt sums up the life work of Father Drum in regard to souls, in retreat or under his direction:

> If the vibrations of our life be Godward—in the service of God our Creator, and for the loving possession of God our Father—they are in order, they are harmonic; our life is music. If the vibrations of life be awayward from God, in discord with His purpose in our life, then life is out of order; it is noise. Like to the G-string is life. The music of a life that is Godward is a deep, full, soul-reaching sweep of sonorous harmonies, an *aria sol G:* grave, gentle, gladsome, grand, graceful, glorious, Godly,—a suite of harmonics on the G-string.
>
> Life out of order is discord, up, down, over and over, back and forth, an ear-breaking noise on noise, griping groan, grilling grumble, grating grouch, gnawing grief, gnarling grunt, growling grind—one long set of noise on noise, a discord and not an *aria sol G.*

With an odd humorousness that may be readily recognized as peculiarly his own, "Drumonian" as he himself would say, at the conferences he would now and again "trot out" his "galaxy of worthies," a few new ones each day, in the form of a description of purely imaginary and exaggerated characters to drive home his lessons. "They are possibilities," he explained, "not necessarily facts." Thus one day it would be the stately Sister Glacialine (pronouned Glacialeen)— the living iceberg, who carried a cold wave wherever she appeared. The thermometer was visibly affected by her presence. Another day the victim would be Sister Niobine, "like Niobe, all tears." Again it was Sister Termagantine, the goddess of war, a "folio edition" of Sister Acidine, all tartness and red fire; or Sister Alkaline "who got the blues, the indigo blues that always lead to the boo-hoos"; or perhaps Sister Saccharine or Sister Innocentine, "who had no defects whatever," and who at the sight of any of the other more formidable worthies sailing towards them in the distance, would instantly disappear, saying: "Yours around the corner"; Sister Apolline, took her name from Apollo, the patron of rain; this was the modern water-sprinkler, always throwing cold water. There were also Sister Mosquetine, the teaser, Sister Gyrovagine, the circulating library, Sister Galline, Sister Gaddine, and other exaggerated types, brought in for the sake of imparting the needed admixture of grave and gay in the hard mental strain of a retreat.

Sometimes the humor was not of his own seeking. For instance, this paragraph is taken from letters to the present writer:

> So glad you were pleased with what I wrote concerning the saintly Father Drum. His memory is held in veneration in our community. Another droll incident, while I think of it. An old nun from Galway, very deaf, and yet anxious to sit up in front, and look up and listen to him, expressed herself this way after the Retreat: "Dear saintly man! how well he is acquainted with my loved County Galway!" He had never mentioned it, but had spoken repeatedly of the familiar Jahweh!
>
> He gave us just four retreats altogether [writes another Religious] when another Father came to give a fifth retreat to the Community, the Superior met him at the door, and said playfully: "You might just as well go home again. No one will listen to you. The Sisters are so full of 'Jahweh' that nothing short of a miracle will get them to listen to anything else." Indeed, his glowing descriptions of Jerusalem and Galilee and Bethlehem were so painted upon our minds and hearts that we were quite spoiled for any other retreats. Our Sisters remember with reverence the learned and cultured Fathers Pardow, Shealy and Drum. May God give eternal rest to their souls.
>
> "Jahweh" means "He Is," [so Father Drum explains in one of the *Pilgrim* expositions of the Psalms], and implies the ever-being essence of the one true God. It is wrongly written "Jehovah"; and is the name of predilection given by God Himself as His own Name unto Moses upon

Mount Horeb, when He said: "Thus shalt thou say to the children of Israel: I *am* sent me to you." Exod. iii, 14. In time the name "He Is" came to have a tenderer meaning and to remind the Israelites of all the shepherd care and all the father's love which Jahweh had had towards "the flock of His pasture." "In the desert, Jahweh thy God bore thee, even as a father bears his little boy, in every journey thou didst make." Deut. i. 31.

To say which of these talks was the most beautiful were an impossible task [writes another]. Undoubtedly, the most appealing was his instruction on the "Lost Sheep," or, as he frequently referred to the subject, the "Little Runaway Lamb." Here, his own beautiful translation of the Twenty-second Psalm was recited as only Father Drum could recite it. Merely to hear him pronounce the word "Jahweh" fired one's soul with the love of God. The sweetness, the tenderness, the forebearance of our dear Lord toward His erring ones was so graphically and forcefully portrayed in this exhortation, that all were in tears. And the folly, the self-sufficiency of the Little Lamb that knew so much better than the Good Shepherd where grew sweet grasses, and where flowed fresh waters, was made so vivid, so realistic, that there was not one of his listeners that might not recognize herself in the self-willed lamb who had grown weary of "the same old pastures, and the same old grasses, and the same old waters!" Then the distant cry of the wolf; and the terror of the foolish sheep, caught fast in the brambles, and feeling the shades of night closing round it. Too late had it learned its mistake. Then, the voice of the Good Shepherd calling His poor little truant for whom He had come searching. The answering cry of the little lost Lamb, overjoyed now, that it recognizes the voice of its Friend. All this, and more, made the conference on the "Lost Sheep" one that shall remain indeliby impressed on our hearts.

The meditation on the Agony in the Garden and the Holy Hour given by him are beyond praise. Having himself visited the very spot where Christ suffered His Bloody Sweat, and examined the very olive trees under which He had agonized, his power to describe these can be imagined. One can never forget the reverence and emotion in his voice when he depicted that midnight scene. The pale, awe-struck moonbeams, struggling through the dark foliage of the gnarled, old olive trees, touched lightly, almost reverently, that Silent Form, prostrate on the ground, bathed

> in His own Blood, weighed down with the sins and crimes of all humanity!
>
> Were one to try to write the many beauties of Father Drum's retreat, still living vividly in our memories, one must needs write a volume. What a pity his beautiful conferences were never written and published! What untold good, what comfort to souls would be effected thereby!

It might be expected that such a treasury of spiritual riches that was being opened to the Religious would not be permitted to disappear with the passing of the retreat; and on almost every occasion the conferences and meditations were taken down very completely and transcribed. A Religious Superior writes:

> It is customary to have the sermons, conferences and instructions typewritten during the free time of the retreat. As Father Drum's were considered of a high order of spirituality and religious perfection, they have been carefully preserved for future reference. They number about two hundred and fifty pages.

On one occasion he noted that there were many pencils and pads in readiness among the audience before him, and was careful to preface the points with the remark: "I have no objection, dear Sisters, to your taking notes, but I have every objection to your publishing them, or putting down what I do not say."

Five or six sets of these retreat notes were in Father Drum's room at the time of his death, but it is probable that many more sets could be available if necessary. A glance through some of them reveals the truth of the statement that "he had only one retreat and always gave the same one," but it by no means follows that the meditations and instructions in these notes are mere repetitions. Though the general outline is the same, the individual meditations and conferences are at times as unlike as if given by different directors. One is steeped in the light of the Old Testament, another is full of allusions and developments of the thought of St. Paul. "I have now an Ignatian, and a Pauline Retreat," he said in a letter, "come and get some Paulinity." All are original in setting and development and in mode of expression. For instance, the Foundation was scarcely ever given in the usual form of a homily on the words of St. Ignatius; but, after explaining clearly the main points of the doctrine that must first be grasped at the outset of the retreat, Father Drum would spend considerable time on a study of the various aspects of God towards the soul, and in every case bring home the relations of Creator and creature in a strikingly original and forceful manner. God is our Shepherd, Father, Architect and Builder, the Husbandman, the Friend, the Saviour. One or more of these aspects are taken for each retreat. Each might form the kernel of a single meditation, and developed from some striking idea in the Old or New Testament, would furnish a ready means for opening out a beautiful exposition of some wonderful Scripture passages, and enforcing the most vivid lessons. The possibility of introducing realistic descriptions of Biblical scenes and places which he had himself visited and studied was patent; but throughout, it was the central thought of the particular meditation in the Exercises to which all these amplifications were applied. Here is his introduction to the first meditation:

> The "First Principle" is for right reason, and the "Foundation" is for strong will. It is a hard and dry principle,— as hard for reason as hardtack is for the stomach; as dry for reason as sea-biscuit is for the stomach; as dry as sea-biscuit and kippered-herring, when there is water, water everywhere, and not a drop to drink." We shall try to soften this first principle. The chief thing is that you chew well on this first principle and get out of it all that there is in it for you. You must become intellectual ruminants these days. I shall present you the cud, you chew on it, ruminate, and make the principle part and parcel of yourselves—flesh of your flesh, bone of your bone, and brawn and sinew of your spiritual life.

The following selections are fair samples of Father Drum's method of developing the meditations of a retreat. The first has been considerably abridged, but conveys with sufficient clearness the lesson of persevering "in the same old way, following the same old Shepherd to the same old grasses, the same old waters," that proved to be perhaps the most moving recollection of all his retreats:

THE GOOD SHEPHERD

First Prelude: Let us consider God as the Good Shepherd of our soul and try to realize His tender care for the flock of His pasture.

Second Prelude: Beg of God the grace to know Him better, in order the more to love Him and follow Him whithersoever He goeth.

First Point: Let us contemplate Almighty God in His relation to our soul as the Good Shepherd. In the Old Testament the fullest description of the attitude of God, the Shepherd, to the soul, the sheep, is that of David's Twenty-second Psalm, written to the God of Israel.

Poet:

Jahweh is my Shepherd;
I have no want!

Sheep:

In pastures of tender grass He setteth me;
Unto still waters He leadeth me;
He turneth me back again;
He guideth me along right paths
For His own Name's sake.
Yea, though I walk through the vale of the shadow of death,
I fear no harm;
For Thou art with me;
Thy bludgeon and Thy staff, they stay me.
For Thou art with me;
In the presence of my foes;
Thou hast anointed my head with oil;
My trough runneth over.

Poet:

Ah, goodness and mercy have followed me
All the days of my life:
I will go back to the house of Jahweh
Even for the length of my days.

Jahweh is my Shepherd;
I have no want.

This psalm was written by David at a time of great strain. In the strain and in the pain of the anguish of those days he wrote: "God is my shepherd, I have no want." Our emotions have their wants and our feelings may have many tendencies, but the wants of the feelings are nowhere, or next to nowhere, by contrast with the wants of the will and the wants of the reason. God, the Shepherd of our soul, is first, and the only want worth while is the spiritual want, which is grace.

In pastures of tender grass He setteth me;
Unto still waters He leadeth me;
He turneth me back again.

Sometimes it is not easy to find tender grasses, especially in the dry season, and so the shepherd brings the sheep higher up where the grasses are tender. We have such tender grasses here in our Religious life, especially in the Novitiate. The tender grasses are God's graces. God's graces are ours in abundance. It is also difficult to find still waters, but in the dash and splash of the torrent the Shepherd knows where there is a turn where the waters are quieter, where the little rills gather into a pool, and thither the Shepherd brings the sheep. God, the Shepherd of our souls, has brought us, in a shepherd's loving care, to the still waters of His grace. Far away are we from the dash and splash of the world and its pleasures. Here in our Religious life we may drink to the full of the still waters of God's graces. The Holy Spirit is never in storm; The Holy Spirit is always in still waters, in the still and the calm and quiet of the soul, when all the world's voice is still. Then can be heard the sweet, gentle voice of the Spirit of Jesus calling to us "Come to me all ye that labor and are heavy burdened and I will give you rest."

Second Point:

He turneth me back again;
He guideth me along right paths
For His own Name's sake."

The shepherd turns the sheep back again because he is the shepherd. He would not be a shepherd if he failed to turn the sheep back when they were near to danger. The shepherd turns the sheep back by means of a staff and by means of his voice. The shepherd knows each sheep by name and calls them by name. Where we love we at least remember the name of the one we love, and so the shepherd knows the name of the sheep and turns the sheep back from danger by calling each by name. So, too, our Superior, the voice of God, the Good Shepherd, turns us back again. Sometimes the word used is not very agreeable, but the voice of our Superior should be agreeable to us; it is the voice of God. The shepherd also turns the little sheep back by means of the staff. The staff is used in two ways by the shepherd. First, for tapping. The shepherd never uses his staff as a sword, he just gives the sheep "love-taps." The staff is also used for stoning. The shepherd picks up stones and flings them at the sheep. He never means to hurt the sheep by the little pebbles that are flung, they are merely to call attention to the danger. So, Almighty God uses the staff of his love on us, sometimes giving us a love-tap, and sometimes giving us a little stoning. The little stones do us good and we must remember the little stones are from the love of God.

Third Point: In the parable of the lost sheep our

Divine Lord portrays Himself as the shepherd, the soul as the little straying sheep. The sheep had grown weary going the same old way, day after day. He thought the grass was greener over yonder, and the waters cooler and clearer. So one day when the ninety-nine turned into the valley, towards the fold, the foolish, restless sheep scampered off. How green the grass! How clear and cool the waters! Oh! it had a pleasant time! But night came on, and the chill air reminded it that there was warmth in the fold. It heard the howl of the wolf, and in its fear, it became entangled and caught in the briars. Nearer, nearer came the howls of the wolf, but,—across the clear cold Syrian air—it hears its own name! The shepherd knows each by name; the sheep knows the shepherd's voice. The shepherd has left the ninety and nine to seek this little wayward sheep. There it is entangled. It can only bleat, bleat, bleat. Nearer and nearer comes the the howl of the wolf, nearer and clearer comes the voice of the shepherd. Which will come first? There is the sheep—trembling, hoping,—bleating, pleading! The shepherd comes. "It is I, fear not." Tenderly he lifts the little thing into his arms—and dares the wolf, who stands at bay. The shepherd bathes and feeds the little thing. We may be sure, after that, this sheep is the nearest and dearest to the shepherd. There is a great deal in this allegory which we may apply to our souls. Our souls would seek their liberty away from the shepherd, away from the ninety and nine. Once the souls stray out into the night of sin, the devil, the wolf, howls his menaces of fear and dread and despair. The brambles of our emotions hold us down, but all the while the Good Shepherd, Our Merciful Lord, is calling, calling. It is dark and cold in the soul when mortal sin is there. But the Divine Shepherd seeks us out, takes us up, washes up and heals us with the oil of absolution and feeds us with His own precious Body and Blood. In return He asks only that we try to be true to Him, and go with *strong will*, the same old way, day after day.

THE GENERAL JUDGMENT

Wisdom v, 18-21. *"He will arm the creature for the revenge of his enemies.......and the whole world shall fight with him against the unwise."*

REHEARSAL

All great plays are often rehearsed before production. The more important the play the more serious are its rehearsals. The more important the act of a play, the more detailed should be its study and preparation. In the play of our little lives, there are no more important acts than those last acts that end with eternity. We should have many a dress-rehearsal. We should go over every detail. We cannot go through all the details of this denouement; we can foresee the general outlines thereof.

The heavy villains have often to rehearse, so as to do their part in proper style and handle their weapons as if they meant business. We shall do our part in proper style. What part will it be? There's the rub! It will be the part that we now choose. Have we chosen the heavy villain's part? Then our end will be villainous at the judgment.

We must think it all out,—think *things*, bring *things* home to us, realize that we shall in the end of time, go to eternity by way of the judgment.

Maybe, in God's Providence, to satisfy divine Justice. *Ut justificeris in sermonibus tuis et vincas cum judicaris.* That thou, Oh just Judge, mayest be seen to have been right in all that Thou hast said to us; and,—we fools,—may be seen by all the world to have been wrong, to have been stupidly wrong in the one thing that should have been right. That Thou mayest conquer as Judge over the foes that judged Thee wrong. So many there are that pass even God before their puny and distorted judgments.

Act 1. Scene 1.

THE CALL.

Matthew, xxiv, 31. He sends ahead his angels. They blow the arousing blasts on their trumpets. They shout aloud the call.

Luke xxi, 34. "Look to yourselves, lest perchance your hearts be overcharged with surfeiting and drunkenness and the cares of this life, and that day come upon you unawares. For like a snare [a noose] shall it come upon all. . . ."

Act 1. Scene 2.

THE DEAD RISE FROM THEIR GRAVES!

I Cor. xv, 52. St. Paul has accurately set this scene before us. "Lo! you now. I tell you a mystery. We shall all rise, but we shall not all be changed." We shall rise as we died. "In a moment, in the twinkling of an eye, at the last trumpet,— for the trumpet shall sound, and the dead will rise again incorrupt."

How many will rise? All. There are now about 1,500,000,000 on earth. After 100 years that number will have died and the same number will have been born. All will rise!

As they rise, all think of God alone, of the Judge, of His judgment! No thought of the creature!

The soul comes back. Dust takes shape again!

Dust returns to flesh-form, not to soul-form. "Dust thou art, to dust returning, was not spoken of the soul."

Act 1. Scene 3.

THEY COME TO JUDGMENT

Matt. xxiv, 31. The angels will gather in the elect (and the damned) from the four winds of heaven.

Matt. xxv, 32. "And all nations shall be gathered together before Him."

See the onward rush and push! All move toward the Valley of Jehosaphat, through which the Cedron flows. To the left are Moriah and Sion, to the right is Olivet. See the grim ranks, the strange procession! 1500 millions of souls today! How many since Adam? See the eternal despair on the faces of some—Judas, Caiphas, Pilate, Nero, Domitian, Julian! Where am I? See the twelve! Judas should have been there! See Paul! Where is Dumas? Where am I?

Pisa, Campo Santo,—building in Tuscan Gothic, 1270. In the quadrangle are fifty-three shiploads of earth brought by archbishop from Jerusalem. Surrounding the quadrangle are beautiful arcades adorned with frescoes. In the fresco of Judgment, the poise and pose of the Judge are wonderful. The face is calm, kind yet determined, the left hand points to the wound in the side, to the pierced and Sacred Heart of Jesus; the right hand is coming down inexorably in eternal damnation. The details are manifold and varied. A monk is coming out of a grave and making for a group of his brothers to the right, an angel nabs him by the hair and pulls him toward the damned left.

Act 2 Scene 1.

THE TRIAL—THE JUDGE

He shall appear! Matt. xxiv, 30. "Then shall appear the sign of the Son of Man in the heavens."

Math. xxv, 31. "The Son of Man will come in His might, and all His angels with Him, and He will sit upon the seat of His majesty."

I Thess. iv, 15. "The Lord Himself shall come—with commandment" at the voice of the archangel, at the blast of the trumpet of God—will come down from heaven.

Just as He foretold it to Caiphas, He shall appear. The day of wrath and no mercy.

See Him now in His wrath, fear Him in His wrath now rather than when the *dies irae, dies illa, solvet saeclum in favilla.*

When Father Perry was sent to see the transit of Venus, he gave accurate instruction to his co-workers, trained them how to see well. The Judge is passing by. See Him.

I Cor. xv, 24. *"Afterwards the end!"* He hath handed over to the Father the kingdom won by his Blood. He hath put down all rule, all power, all might. He must reign, till he hath put all enemies under His feet. The last enemy he destroys is death. *Deinde finis!* Sublime passage! All power is taken away. No more power in demons. No more power of grace. The end! All power is His! How will He use that power? In justice!

The judge without mercy! Hebrews x, 26. "For if we sin after that we have received the knowledge of the truth, there remaineth no more sacrifice for sin; but only a certain doom of judgment and the yearning of the fire which is going to devour the foe." There was no mercy for him that was proven by two or three witnesses to have despised the law of Moses. "What sorer punishment shall he, think ye, be deemed worthy of, who hath trodden under foot the Son of God, and hath counted as common the blood of the testament wherewith he was made holy and hath done insult to the Spirit of grace?"

Dread text! No prayer! No sacrifice! The Blood of the Saviour has too long been held as commonplace. No Mass more! No more treading upon the Son of God! The witnesses are there! The books will be read. The books tell truth. Then you are damned!

All the rest of life, there was sacrifice for sins, there was blood and enough,—more than enough of blood, not common blood.

The Judge! Not the *mitis et festivus Jesus* of St. Bernard; rather *rex tremendae majestatis, qui salvandos salvas gratis,"*

Lord of dread and Lord of Might
Who freely saves the helpless wight!

God's Justice

Don Alvaro held Diego, Pereira at Malacca, and opposed Xavier's journey to China. On his departure, Xavier refused the request of Dr. Suarez (the bishop's vicar) nor would see Don Alvaro; but shook from his shoes the dust of Malacca, prayed for Don Alvaro, and sent the message: "Don Alvaro, I shall never speak again in this life. We shall next meet in the valley of Jehosaphat." After Xavier's death on San Chan, Don Alvaro's men wrote him that St. Francis had died like any other man, and had not worked wonders before death. These men turned coat on the way home to Malacca with the precious remains of Xavier.

THE BOOK

"The book shall be opened". . . . (Apoc. xx, 12).

Whether from winds of heaven or dust of earth, every particle that was made flesh will be summoned back. No matter how many chemical changes will have taken place since our flesh and blood will have been dissipated into gases and other results of death, that flesh and blood will once again be mine, will be informed and vivified by my soul; and I, this very I, shall be judged in flesh and blood, soul and personality, what time "the book shall be opened to all men." To all men shall it be opened; and they that canonized me (with one "n" for a time, will cannonize me (with two "n's") for an eternity. "The whole world will do battle for Him against the fools!"

Act 2 Scene 2.

(a) THE JUDGE

Mt. xxiv, 30. "They shall see the Son of Man coming in the clouds of heaven with great power and might." Just as Christ foretold to Caiphas.

Mt. xxvi, 54. Caiphas had asked: "Art Thou the Christ? "Thou hast said it,"—i.e., "I am." Maybe there was then a jeering laugh of scorn and scoff! "A King? You look the part."

Verumtamen! None the less! I may look like somebody else. I may seem to be a mountebank. You may seem to be the judge, I the adjudged! You are seated; I am standing. Things seem your way, not my way. *None the less!* The time will come when I will sit and you shall stand; I will be your Judge and you shall be adjudged. "You will see the Son of Man seated at the right hand of the power of God and coming in the clouds of Heaven." Caiphas should have understood. He did understand. He tore his garments, for that Christ had blasphemed. Caiphas shall see the Judge! They all shall see! The adjudged shall see!

Act. xvii, 18. The men that made profession of, and had reputation of being thinkers;—i.e., the Epicureans and Stoics of Athens poohpoohed St. Paul's words. *Quid vult seminiverbius hic dicere?*—picker-up of seeds, a lounger, a beggarly fellow, an empty talker, a *babbler*. "Why he seems to be a peddler (a hawker) of strange gods!" Before these skeptics and scoffers and Epicurean deifiers of pleasures of sense and Stoic deifiers of the supremacy of spirit over matter, St. Paul made bold to preach the resurrection and the after-life. He boldly met a great difficulty face to face. Since we are God's own race, how is it we have gone wrong? We have made gold and silver and stone, the handiwork of the art and thought of man, to be like unto God! How have we gone so far wrong, if wrong at all? It was a hard nut for Paul to crack. He cracked it. "God hath winked!" [The force of the Greek word is not expressed by *despiciens.*] "God hath *overlooked* [taken no notice of, winked at] these times of ignorance." Now, not so! He now announces to you the truth. He has set a day. On that day He will judge the world in equity, in the [name of the] Man whom He decreed, the faith in whom He hath offered unto all, whom (in fine) He hath raised from the dead! What was the result of this frank statement of Paul?

Some laughed at him. They shall see! Some said: "We shall see you again!" They saw no more! They shall see! The third class believed. They shall see! All shall see the Judge on that last day!

Act 2. Scene 2.

(b) SELF

They Shall See Themselves!

At first all will try to excuse themselves. The lecherous: "I could not help it!" "Did you use your will-power? Did you draw upon grace-power by Communion? No? Then you did not try to help it! Then you are damned!" They shall see themselves! All will be dreadfully clear,—clear-cut and standing out in bold relief to the gaze of everybody. They shall see them! Why not see now? Borgia meditated on these first week truths five hours a day for ten years. He saw! He saw and judged from the standpoint of the judgment.

When we, in very truth, see ourselves on Judgment day, gone will be the standards of human respect and fear and lust and curiosity and pride; gone the opinions colored by flesh-love and world-love; gone the notions perverted by vanity.

No appeal! The one judgment of one Judge is forever and final. No appeal to bad example, to lack of light, to human weakness.

ANSWER

What Have You To Say?

The lecher writhes and twists. What have you to say? To say? There was a time when he thought, with Talleyrand, that speech was given to man as a mask for thought. That time is past. All masks are down. The masquerade of life is over. The white light of eternal truth shines out. Every man's soul stands clear and in bold relief.

The infidel and man of no faith: "I wronged no man." "You wronged Me!" "How is that?" "You hated Me!" "Why I never thought of you." "Precisely! You ignored me. Now, I ignore you. You got along without Me. I will get along without you. Your mother taught you better. You were impure and proud. You lost your faith by pride that kept you from confession. To the left!"

OTHERS

What surprises! The penitent among the elect sees his confessor among the reprobate! The pupil saved sees his professor lost! The railroad president on the left sees the railroad gang on the right! The capitalist is not with the toiler, whose life-blood he coined by a miserable wage.

Act 2. Scene 3.

THEY SHALL WEEP

The man of millions shall weep, the man of brains shall weep, the man of toil shall weep, if he be in sin. All his millions, all his brains and all his toil shall the sinner wish to give up for one saint to plead his cause. No pleading! No saint! Only sinners! They shall weep! Caiphas shall weep. Judas shall weep! The villain who took out his watch and calmly and cooly counted out five minutes in which time he dared God to strike him dead,—that clown, who took it the Creator should be a mountebank,—*he shall weep!* God is not at the beck of every knave; He is no second-rate actor to come upon the stage at the atheistic push-button's signal. He will come then! He had said: "Blessed are they that mourn!" He had canonized sorrow! Only in this life! Not on that day! Sorrow now is as nothing by the side of sorrow then. Sorrow then will be for sin seen in all its iniquity. . . *"Si iniquitates observaveris, quis sustinebit?"*

Act 3. Scene 1.

SUMMING UP!

Hebrews x. 30. *"Mihi vindicta et ego retribuam!"* Vengeance is mine and I shall pay back. For the Lord will judge His people. *Horrendum est incidere in manus Dei viventis!*

Act 3. Scene 2.

Luke xxi, 36. "Watch ye, therefore, and pray at all times that. . .ye may be accounted worthy to *stand* before the Son of Man"—worthy to receive from Him a sentence of worth.

Act 3. Scene 3.

Wisdom v, 18. "His zeal will take armor and he will *arm the creature* to take vengeance on the foe."

Romans viii, 21, 22. The creature groans. "Lord, let me open up and swallow him."

The *armor* for execution: Wisdom v, 19—*Breastplate of justice.* No loving tenderness. No mercy. The cold steel of justice girds the warm love of the Sacred Heart. That Heart may no longer be reached. No *ora pro me!* No *miserere nobis.* Even the Blessed Mother could not reach!

Helmet.—true judgment, certain knowledge. He sees all that is in man, and all that is not!

Redoubtable shield, equity. Even the prayer of our Blessed Mother would rattle against that shield and fall to the ground. Cries? Tears? No effect! Equity! Fitness! Prayer not fitting! No more! Did prayer pass shield, it would never pierce the cuirass! Prayer is out of place on that day—even prayer in grace! Will no prayer be heard that day? Yes, the prayer of justice! It is the day of justice. The prayer for justice will be heard! Against the man who has given scandal, who has dragged the innocent to shame, the soul of the damned will cry for justice: "Lord, that man drove me here! Deal justice out to him." The horror of it!

Lance.—"He will sharpen to a lance [point the hard metal of] His wrath." Day of wrath. You may pray for wrath. He will be wroth! While on earth, He was angry only at the hypocrite, the Pharisee; never at the Magdalene, never at the poor woman taken in sin, never at Peter, never at Judas. In the Blessed Sacrament, He is angry at none! On that day, He will be angry at every sinner! On that day for the first time, we shall see Christ. We fancy Him in sweet calm and loving repose. Shall our first sight of Him be the sight of the Christ in anger? Shall we feel His hard wrath pierce us through as a sharp lance? Shall we be against Him when He has armed the creature to fight against the fools. Is this the Christ? The *mitis et festivus* Jesus of St. Bernard? This is the Christ, the Judge of the fools! Then we must tear up the pictures that we love of Him. Perugino's Crucifixion, with its tender and soulful love-gaze of the Christ at the Blessed Mother—tear that up! All the many drawings and studies and paintings that Guido Reni made of the thorn-crowned head, those noble, pleading eyes that show so great a love of reparation—tear them up, every one of them! We will have *truth!* There is the meek and gentle Sacred Heart, flowing over with love for all! Is that false? On this day, false! Tear it up! There is the dear Saviour on the cross! Tear that up! We may look at Christ in His mercy no more! We must look at Him today only in His justice. No artist has ever done Him justice so; yet so will He be on the day of justice and of wrath!

As I gaze and ponder, I beg and pray. Lord, throw away that lance; throw off that helmet and let me see Thy loving eyes; oh, take off that cold, steel breastplate and throw away that hard shield. Let my cry reach Thine ears. Let my heartbeats reach Thy heart. Let me see that Thou really hast a Sacred Heart! What! Can it be? The shield is thrown away! The lance is dropped! The helmet disappears! The buckler bursts! The tender eyes look love into mine. The Sacred Heart I see again. The sacred hand beckons. The sacred voice calls. Can it be? Is it all true? It is, dear child, that was only a rehearsal!

MEDITATION ON THE LOVE OF GOD.

I. HAPPINESS.

We should always be happy, *Gaudete semper,* "be ever glad!" I do not know whether I told you of the notion that our dear Saviour never laughed. This is apocryphal, absurd, unscientific. Our Lord was a perfect man, and a man with all the perfections of a man has a sense of humor. The idea that our dear Lord never smiled is to be traced to the decription given of Him in an absurd, apocryphal document called the "Letter of Lentulus."

II. PURE LOVE.

Why should we be ever glad? Because we have pure love of God. The dear Lord is in His glory, the Heavenly Father is infinitely happy. Then I am glad! My motives of sadness are self and in the interests of self. No such motives are in God.

(1) A very beautiful example of the love of friendship which our dear Saviour proposes to us, or rather

which St. John the Baptist proposes to us, in regard to our Saviour, is the love that the friend of the bridegroom has. For a moment you stand aside from the opinion that Christ is the mystical Spouse of the soul of the virgin, and take a larger view which makes Christ the mystical Spouse of the Church. Both these views are correct.

(2) Another good example is that of a wife for her husband. We often see this in our priestly ministrations. It is the highest form of human love, this love of self-sacrifice, of a wife for her husband. I do not refer to the passion of love, but to the love of self-sacrifice a Catholic wife will show her husband for twenty-five or thirty years or more! Such pure love of friendship is not necessarily in the pasion of love; quite the contrary; it is founded on reason and not on passion. Why such absolute sacrifice without the slightest murmur of regret? For pure love. The wife is satisfied as long as her husband loves her. He comes home after a day of hard toil, smiles, is sincerely pleased and gives loving approval. That is enough. They are absolutely happy. If, however, the husband is cranky and shows a lack of affection, the wife broods and is sad in her pure and self-sacrificing love for him. There is a oneness in their lives, aspirations, hopes, disappointments, successes, failures. Such should be the love of the soul for Christ.

(3) A third example is the human love of a mother for her child. We priests sometimes meet great poverty and even suffering that women of the world bear heroically. The motive that stays them is at times pure love of a boy, maybe a wayward boy, maybe a boy deserving of a mother's love. Instance: "Aren't things too hard for you to bear?" "Oh, no, Father, I have a great deal to make me happy. Why, my boy will be a priest this year! What more do I want?" We should bear it in mind that our dear mother's joy was often such, a joy of pure love for us. Such should be our joy of pure love of God.

III. Love is Shown in Deeds.

The meditation we shall make is called the "Contemplation to Obtain the Love of God." Père de Ravignan said in his last retreat that the love of God does not consist entirely in the dispositions or tenderness of the soul. These may form a portion of love, it is true, and may tend to the love of God; but do not constitute it. There is *need of deeds* with the motive of pleasing God. There is need of practicing virtue even unto perfection. It is true love that makes us always turn to God as our sovereign

Good. St. Ignatius puts this in the form of two principles that precede his meditation.

St. John, in his first letter, chapter three, says: "Little children let us love not in word nor with the tongue of language, but in deed and in truth." St. Gregory puts the same truth in these words: "The proof of love is doing." St. Ignatius lays down two principles: Love shows itself in deeds rather than in words; Secondly, love shows itself in a communication, so far as that communication is possible, of the good things of the lover to be beloved, and a return of the good things of the beloved to the lover. That is love in every form, in human love and in Divine love.

Now we are going to study how God has shown His love for us by giving us of the good things of His Divinity, so far as it is possible for the Divinity to share with us. Before we go on I wish to call your attention to an instance of love. It is the love that St. Augustine had for God. That love is described in his "Confessions," in one of the most beautiful and touching chapters. In this description, St. Augustine is about to tell of the death of his mother, St. Monica, and by the way, that description is wonderful. They are on their way to the African shore; the two sons are very much concerned; they think they will not reach the shore of Africa in order to allow their dear mother to die there. She says, the dear saint that she was: "Lay my body where you will, but when you stand at the altar of God, my boy, think of my soul." On this occasion both of them set out for Ostia on the Tiber. Augustine and Monica are in an upper chamber of a building that overlooks the Mediterranean. They go over to the sill and gaze out upon the sea.

Referring to his mother again in the tenth chapter, Augustine writes: "Not long before her death, she and I stood together and leaned out over a window-sill so as to get a glimpse of the inner garden of the house. We were talking, oh, so sweetly. We had forgotten all the past for all the future that was before us. We were wondering what Thou art, O God, and what should be the life everlasting of the saints in the hereafter—that life which eye hath never seen, nor ear ever heard tell, nor hath it ever entered into the mind of man to know. As we stood there with hearts agape, we drank in deep draughts of the heavenly waters of Thy fountains.—of the font of life with Thee."

They then rose higher in talk, until their thoughts were beyond expression. They remained fixed in ecstatic gaze upon the wonders of the Godhead and upon eternity where there is no *has been* nor *will be,* but all *that is.* In that wonderful contemplation, they both passed far and away beyond the human word, and pierced in thought nearer and nearer to the Word Divine.

After this sublime meditation, Monica told Augustine that the things of earth palled on her; she cared for them no more. "As for me, my son, there is now nothing on earth I care for. I know not what to do here now, nor why I stay here longer; the hopes I had in this life are come to pass. There was only one reason why I used to wish to stay in this life a little longer. I wished to see you a Christian and a Catholic before I should die. God has granted me my wish and more; so that now I see you His servant and a man who despises the things of earth."

About five days later she was taken ill with a fever. One day she became unconscious. In coming to, she saw my brother and me standing by and said: "Where was I?" She marked our grief, and went on: "You will lay your mother away here." I said nothing and could do no more than keep back my tears. My brother said he hoped she would die at home and not far away from home. She scolded him with a glance for thinking such a thing and then looked at me. "Hear what he says! You will lay this body anywhere; have no bother about the care of it. I asked of you both only this—that wherever you be, you remember me at the altar of God." She did not speak again.

The meeting of Augustine and Monica is beautifully painted by Ary Scheffer. This scene is one of the few modern paintings (1850), which held my rapturous attention in the Louvre. Monica and Augustine at the window-sill! Monica, clad in pure white, looks into vast heaven. She is seated on a low stool. At her feet and seated on the ground is Augustine. Their hands are clasped. Their hearts are one in sentiment. Their thoughts are one. *De longinquo te aspicio!* Monica taught her wayward boy to look to God. They are gazing out upon the heavens. They talked not of all his sinful days, nor of all his dear mother's prayers. They talked and thought of God alone as they gazed out beyond the stars at Thee, dear Lord, far, far away! From far, far away they gazed at thee, and talked over all Thy love and grace. They gazed at Thee, and talked and gazed; and talked and gazed and gazed and talked no more! They were rapt in an ecstacy of love—that dear mother and her wayward son! Now we wish to make an *act of perfect love like that.*

POINT 1.

In the first point, we consider all the gifts God has given to us. I think of all the gifts of *inanimate* nature, all for *me!* Think of the rocks which furnish the solid bases of buildings; the waters holding the fish; again, the waters of the sea and of the lakes as a means of navigation. Think of the motor power, electric power, sound power —all wonderful powers of the physical world, discovered by man but made by God. Think of the forces of creation; of the cohesive and repulsive forces in nature, causing atomic affinities, exercising molecular motion, thus giving us light and heat and electricity. Think of the myriads and myriads of microscopic particles that make up the molecule in nature!

Then consider *animate* nature. Think of *plant life*, the flowers in all their beauty—the lily for instance. Think of the plant life that is not merely beautiful to look upon but useful as food; the gifts of all the cereals, wheat, corn, grain of so many sorts. Then we have the vegetables, herbs, trees—a wonderful variety of fruit trees, all for *me!* All these are the gifts of God's love for me!

I then come to the order of *animal life*. Animals are very useful to man; they provide food, pleasure and give means of transportation. Then, from the order of inanimate nature, animate nature, animal life, I rise to *men* and think of the gifts to me from God's love: my father, my mother, all my family and my relatives; of the men and women I have met who have been, indeed, gifts to me! I then rise from the order of nature to the *supernatural* order and see the gifts there: non-sacramental grace, sacramental grace, the myriads and myriads of flashes of God's light into my reason, the myriads and myriads of spurs to my will to do good. He has given me all these: actual grace, non-sacramental grace, sacramental grace, the gifts in the sacramental life of the Church, the communions for myself, my confessions, the absolutions He has allowed me to receive, the communions He has allowed me to receive.

Realizing the love that has prompted all these gifts, then I ask myself: *"What shall I give back to God, for all God has given to me?"* I then say my prayer—*Suscipe!* "*Take* and *receive*, O Lord, all my liberty, my memory, my reason, my will, all the powers of my soul. All that I am and all that I have—*take* them! They are Thine! All that I have and all that I am are Thine by right; Thou art my Master. In all justice, I beg Thee, take them all; I give them back to Thee; by every right they are Thine. But, O dear Lord, even though they were not

Thine by every right, even though Thou couldst not and wouldst not take them in justice, yet I would give them back to Thee. *Receive* all that I am and all that I have. With every reason I offer all to Thee, because I owe all to Thee in return for all that Thou hast given to me!" Say this prayer in your own words. It is not necessary to say it with emotion; say it with reason and justice. Sanctity is a matter of *reason* and of *justice*, and not an unreasoning emotion with no sense of justice.

The full power of generosity is now turned on after the logical work we have done. There is no drag or hindrance to the logical conclusion. One has full fling to offer oneself wholly to the dear Lord. There is nothing of the child in this offering. There should be the manly love of St. Paul, who said: "When I was a child I spoke as a child. . . . Now I am a man and have thrown away entirely the things of a child." (1 Cor. xiii, 11). It won't do for a man to be a baby when he is called to battle. The soldier who does the baby-act is despised—gets his mother to swear he is under age. The battle-call is on! The foe is advancing, and now in strong manly-wise, in strong womanly-wise, in the way of the valiant Judith, let us strike the foe down.

Sume! Take! "It is all yours, O Lord, to do as you will." Do I really mean to give my liberty to the Lord? "No, Lord, I may as well be honest with Thee. I mean to keep my liberty! I know Thou canst take it against my will, but I am not going to offer it. So long as Thou dost not use Thy eminent domain over my rights and privileges—my liberty, my sweet will, then just so long I am going to talk as I please, to go with whom I please, to act just as I please, to carp at any person whomsoever, to read what I please!" Oh, then, I shall rattle off that beautiful prayer of St. Ignatius. It does me good to fly high, at least in words, but those words mean only what I should like to be. Oh, no, I shall not be a poll-parrot in my prayer; rather let me omit the *Suscipe*. Omit the *Suscipe?* That would never do! And yet the *Suscipe* is *only one step* to the *act of the Love of God*.

Conclusion: Thy love and Thy grace give me: These alone are enough for me. Do Thy love and Thy grace satisfy me, dear Lord? I wonder! Not by any means! I wish to be praised, to be thought well of, to come to my own!" The conclusion to this prayer is necessary; all is God's. All should be rendered back to God. There is nothing I have any right to retain from Him. Yes, there is one thing to which I have a right—there are two things to which I have a right from God; they are indispensable. I will not give them back to Thee, dear Lord! Thou wouldst not have them back: Thou wouldst hate me for giving them back. These two things are Thy love and Thy grace. They are enough! These are all that are left to me. I am criticized? By whom? By man! By woman! But God loves me: that's enough for me! I am misunderstood! By whom? By man or woman! No matter, the dear Lord loves me and that's enough! But my Superior wrongs me; she at least should right me! She is a woman—Jesus loves me! I have enough. The dear Lord knows me through and through and He says to me: "That is *well done;* that is done *for Me!*" What more do I want? What more do I wish in life with such an approval? "Thou art such and so great as thou art before God and not before men" (Imitation).

And so I keep *love and grace,* and all things else I give; things of body and things of soul, external gifts and internal gifts, perfections and imperfections, whether of body or soul. Such an imperfection is my incapacity in study, my inferiority in the Community, my work for outsiders. In regard to these imperfections, it is harder than in regard to perfections to say: "Receive them, dear Lord; with reason and justice I give them to Thee."

I make up my mind to get along with only two things: God's love and grace. Shall I hold back some things so as to fill out the deficit in the event of God's neglecting me? No, there will be no such deficit. God will not be outdone in generosity. The more we give to Him, so much the more will He give us of His love and His grace. Love and grace, first, are enough for me; secondly, are mine in abundance. All I have to do is to ask and I have the grace asked for. Do I wish to become a saint after this retreat? Do I wish to be able to make that act of failure? Yes? Then, I ask the grace. Only a child prays: God, make me good. Men and woman come down to detailed facts. I see that this temptation is keeping me from sanctity. I pray for grace to avoid that temptation. I see that vanity is leading me away from Christ-love; that I pray to be humiliated in this or that. Be concrete. Ask to be humiliated in this or that particular occasion or circumstance. The grace is straightway mine, and will mean actual graces necessary for various occasions of such temptations. Then I go straight on as if the grace were provided. Of course the will must work with grace. Grace is not an exterior force working the will, as the foot impacts the foot-ball. Grace is an interior force which raises the will though never against the will, to a higher level, to the level of salutary acts of deeds done meritorious for salvation. Let us remember our Lord's words: "Thus far ye have not asked me for anything." Maybe you have been asking for stones, for serpents, for scorpions. You have not asked for fish and bread and sheep, or, if you have, you have not asked properly.

POINT II.

We see *God present in all nature*, inanimate and animate, in the sacramental gifts, in the external possession of Him by our intellect and will. He gives Himself to us unto the utmost limit of our receiving power. The philosophical principle is: "Whatever is received by one is received according to the limit of one's receiving-power"—potency—and only our limited receiving-power limits God's putting Himself into that which He gives us. The woman who loves may not give more than her service; she cannot give herself in each act of service of her beloved. God gives Himself in each sign of His love.

In the first point we considered the gifts as *sent* to us. In the second point we consider the gifts as *brought* to us by the giver. A gift to a royal personage is not sent by freight nor even by express, as it might reach some lackey. Personal delivery is the sign of respect and love. So God Himself brings His gifts to us, and in them gives us Himself. The dear Lord has loved me so much as to have *sent* me none of His gifts but to have brought them all in person, even to have given Himself in each of them. This presence is by His essence, omnipotence, intelligence, and by His Eucharistic presence. Hence, logically, I should not love this friend because he pleases me, but, because the Lord has given me Himself present in this friend, I should love him with Christian love. I should not stop short in my love, but should go beyond the person to God who is present in that person. I am a baby in the spiritual life if I want other people to stop short at me in their love for me and not go beyond to God in me.

It would be illogical to say: "Lord, your love is too small for me. I want some greater love, something more get-at-able." No, I can't say that to my friend. Again, I should now rejoice at the greater ability of others. Why? Because it is God present in them that gives me joy. There are many in whom I see more of God than there is in me,—more of grace, more of love, more of the spirit of the institute. Then, let me rejoice, because that is God present in them. How wrong it would be to belittle the efforts of others: "Oh, she's nothing much!" or, "She is putting it on!" Then I proceed to pull her down to the dead level of mediocrity. I should, on the contrary, thank God present in her and help her up. Am I ready to thank God for His presence in His creatures to such an extent as to make the *better offering?* Not yet! Yes, I am. I shall with God's grace, lower self, thank Him for the better results of His presence in others than in myself.

And so I offer to God in this *better offering* of the second point all that I have, my reason, my will, my memory, all that I am, and I offer Him myself present in all that is His and mine. I offer Him all things and with greater affection than ever before. I see Him present in all that I offer Him,— in the Blessed Sacrament, in my brethren, in my Superiors, in the world with which I have to do. I promise to see in my Sisters the temple of the Most High. I will not desecrate that temple. I offer myself more fully present in my work, in my prayers, in my Mass, my office, my examens. I not only give my liberty, but I shall see God in every bidding of my Superiors, see God in the judgment of my Superiors, see God in my Superior's will.

Work it out for yourself, Sisters. Make the better offering. Saint Ignatius here wishes us to work out our own thoughts. Offer yourself fully in your vocation. Make the offering of accepting God fully in your Superior. That is even a better prayer than the *Suscipe*. It is coming down to details, and that is important. "When I was a child, I prayed as a child; but now I am a man, I pray as a man."

POINT III.

I see God *toiling in all His gifts for me*. Not only giving to me these wondrous gifts of nature, inanimate and animate, natural and supernatural, not only present in all these gifts, but toiling in them all. Like a skilled laborer, He toils in inanimate nature. It is He that is toiling and effecting the pull of gravity; it is He that exercises in the smallest particle of matter its forces of repulsion, of cohesion and tension. It is He that is toiling in all those forces of inanimate nature, toiling in all the atomic affinities, that bring certain elements together and unite them chemically. It is He that is toiling in the affinities that prevent nature from going awry. You see Him toiling by His omnipotence in plant life, giving being and life, toiling in every cell, causing cell to break up into cell. He paints the lily as an artist, nay, as no artist could! He perfumes the violets. He gives myriads of colors to all the flowers in the field, making the trees to burst and bud in the Springtime, to bloom and bring forth their manifold fruits, upraising the slender spear of wheat or grass, protecting and toiling with it until maturity.

God's wondrous toil is in all animal life, in the production of the wondrous species of animal life, in the perfection of the works of animal instinct. Take the maternal instinct. It works marvels while the offspring is in need. The animal will never fail in its maternal instinct. Woman may fail by perversity of will which is free; the animal has no free will to fail, her instinct is necessary. God toils in it. What God does for a purpose is excellent and sure of accomplishment far and away beyond that which man does. Take the sting of the bee. It is a minute lance, so sharp and penetrating as to defy man's imitation. God toils perfectly for His purpose, the defense of the otherwise defenseless bee. Take the ovipositor of the ichneumon fly. With the very slight force the fly has at its command, she may push his perfect, almost frictionless needle into the hardest wood, more readily than a strong man could push a needle thereinto. And reasoning, in this one line, better by God's reasoning than could man by his own, the ichneumon fly discovers below the surface of the wood, the larve into which she drives her ovipositor and in which she lays the egg. God toils in all this to bring about the equilibrium in nature. God toils in the weaving of the spider's web, a lacework that shows the instinct of the spider to be in this line more capable than the reason of man. And why? Because God toils in that which the spider does!

God's wondrous toil conserves and concurs with all human action. I cannot lift my little finger but God lifts it with me. Every little cell in my body has its almost perfect life, and God it is that is toiling in that little cell.

Going above nature, I see God toiling by grace, toiling by actual grace, toiling in the Church; and I see God Incarnate toiling in His public life, Jesus toiling in His Passion, His risen life, toiling in His sacramental life, toiling in every Mass that is said, toiling in His Eucharistic presence.

What is the *better offering* that I give in turn? What shall I give to God in return for all God has given to me? Toil, toil? Toil, toil! that is the offering—toil in self. He has put powers into me that I may use them. How have I used them? As if they were given to me for my own benefit? As if they were mine for the sake of Him who toiled within me? Maybe I have been all wrong—the dear Lord is toiling in my Sister. Am I trying to make her a traitor? Shall my actions say: "Lord, because I love Thee so much, I shall do my level best against Thee? I shall give wrong notions to my Sisters, I shall try to haul them down to my level." Shall I thank God that I have spoiled the work of the Almighty? No, no, no, let me not go so far even in words. Let me thank God for all that he does in my Sisters, and all that they do for Him by His working within them.

So I offer myself, fully toiling in all that I have to do, and I offer myself to God toiling in all my Sisters, *and I ask God that I may fail,* so far as toil is for myself, and that my toil may succeed only in so much as it is for Him. And now I have realized to the full God's gifts to me and my accepting His presence in all my toil and gifts to Him; God's energy expended for me and my energies that should be expended for him.

POINT IV.

In the fourth point I rise from all thought of myself to the thought of Him. I stand at the window with Augustine and Monica and gaze out in imagination beyond the sea, beyond the farthest star. The sun is 93,-000,000 miles away, and the next star, alpha centauri, is 275,000 times 93 million miles away, and the next fixed star is 576,000 times 275,000 times 93 million miles away. There are stars which have not been reached by the strongest telescope and I gaze beyond those millions and billions of stars, and think of Him who toils in them, and I leap in imagination beyond the shining of the utmost stars. I forget my meanness, O my God, and leap in imagination to Thy infinite Mercy; I forget my contempt and leap in imagination to Thy infinite Truth; I forget my utter lack of goodness and leap in imagination to Thy infinite goodness. I forget all my little, miserable attributes, which are only the faintest reflection of Thy infinite attributes. I gaze, and gaze, and gaze, at Thee, Oh, my God! Thinking of Thee alone, of Thy goodness only, of Thy justice only, of Thy mercy, I make my act of love. Oh, my God, I love Thee with all the powers of my soul, not because of this goodness of which I have been the gainer, but because *Thou art in Thyself infinitely good and worthy of love.*

CHAPTER XVIII

SPIRITUAL DIRECTION

The letters of Father Drum to Religious and to seculars who put themselves under his direction might possibly fill several volumes. Indeed the interest that attaches to the following selections is chiefly that of wonder at the marvelous patience and solicitude they manifest, to say nothing of the phenomenon that he was able to accomplish the work of so much personal and painstaking spiritual direction, in the midst of all his other duties of preaching, writing, and the giving of retreats. One and all evince a genuine fatherly solicitude for all the spiritual concerns, indeed, for all the concerns, of those who had once his friendship, or even had but once made a retreat under his direction.

Obviously there is no measuring the amount of good that was accomplished by this literary apostolate. Only the recording angel, and the many souls who felt themselves privileged to win his devoted friendship, may know its value. But it is enough to note that he kept a careful list of the names of all who, under his personal direction, were led to follow out a religious vocation. The list totals eighty-five Religious whom, from 1908 to 1921, he had directed to the Order of Notre Dame of Namur, the Sisters of Mercy, the Ursulines, and other Religious Congregations; and does not include the probably larger number who had merely sought his advice about their vocation. The spirit that could take time enough and expend energy enough for the patient accomplishment of such an apostolic task, must indeed have been steeped in the zeal and charity he so constantly preached to others. St. Paul's own words "the charity of Christ presseth us," are probably the best commentary on the following selections from his letters. Some of these have been considerably curtailed, for lack of space, and in some cases only such passages have been selected as indicate a work of zeal or an incident that has elsewhere been omitted.

Written to a Religious who in the beginning had found great

difficulty with her vocation, especially on the score of obedience:

There is only one thing to do—follow the advice of one who has known the religious life and has lived in it, one who has experience in the direction of nuns. . . . Give over the plan of making an order to order. Take things as they are: they are excellent. Follow the judgment of your Provincial: give up your own sweet whims. If you follow my urging you will finally be happy as an. . . ,if not, there is no telling what more will come your way of disappointment and sorrow. . . . You are a Religious first for the perfection of your own soul and then for the salvation of souls in Alaska or elsewhere. The latter depends upon your Superiors. Your perfection is your work and God's. . . . In the sweet Providence of God, your experiences may bring you a final joy in doing the will of another. . . . There are human sides to all orders. The main spring of your troubles is not in the order but in your self. Prudence dictates, imperatively and vehemently dictates, that you have direction and follow it. I extend you a strong hand across the continent. If you grasp firm, I shall, by God's grace, swish you into a good Novitiate and give you the opportunity of a Religious training. You have groped enough. Now be satisfied to be told just what to do.

You have suffered enough by wrong direction. And yet there is a sweet and fatherly Providence that allows the mistakes of our judgment and the mistakes of the judgment of others; to disallow them would call for a miraculous intervention. Then after the mistake has been made, that same loving Providence provides many graces to make good the seeming evil,—indeed, if we be true, prudent, and Christly, even greater good comes than might have been, had not the mistake been made. And so, be brave, prudent, patient, full of faith. Read I Corinthians xiii, —especially verse 5: in the Greek of St. Paul, it is: "All things she excuses, all things believes, all things she hopes, all things endures. Love never fails." You will come out of this a happy child of Christ Jesus. . . . You must first become a good Religious: and then do mission work. My sincere blessing to you. Your needs are with me at the dear altar of God.

Your training in a regular Community will make you ready to meet with equanimity such conditions as are in store for your future life in Religion: and to realize the workings of grace in circumstances not at all to your liking. In the years to be, you will gradually come to know how elusive, protean, kaleidoscopic, is self-love. It is overcome in one form,—that of love of ease and home,—and then takes another shape: and so on to the end of the chapter of each little life. May God give you light to see, and strength to grapple with this manifold and multiform enemy of Christ-love. There is a sweet Providence in all your trials. "Trials beget grit: grit begets reliability: reliability begets hope: hope never disappoints."

No, you are absolutely sincere, have been so all along. Do not doubt that; and never doubt your vocation to the order. Ignore all such doubts as coming from the Evil One. . . . He has failed. The grace of Christ, and His sweet love, have triumphed. . . . So, hold firm to your calling. Never doubt it for a willing moment, no matter how many unwilling doubts flit through your storm-tossed fancy. Just think of Him who said, "I have conquered the world!" Then say, "Dear Jesus, You are worth my while!" And go on in the beaten rut, no matter how unappealing it may chance to be. There is no fun in following one's vocation, but there is in the end, a joy past all telling. You will be unspeakably happy. That I promise you. It must be so!

It is a great joy that now at last you radiate peace and love. Take it not ill that your experiences have been painful. A kind and fatherly love has allowed all your trials. They were brought on by the natural sequence of cause and effect. . . . He has brought good out of the seeming evil: and will bring still greater good to your soul in the years to be. Have no regret at now and again having given me pain. It was good for my soul. You knew my intention was your soul's happiness. . . . My determination was, by the grace of God, to see you through just so far forth as you would allow me to do so. You are with me in prayer, especially at the dear altar of God. . . .Frankly, I think that you needed this experience. . . .You needed to learn by mistakes that you should rely on the decision of those who are to you in the stead of Christ Jesus, the Saviour of us, His willing slaves. Your will has been ever beautiful in its devotedness to Him. That is why He has saved you in your troubles. And the experiences of the past have taught you a lesson of obedeience for the future. If I err in this writing, you will graciously forgive: you will take the humiliation as a means to the future sanctification of the soul in which I am sincerely interested.

You have not been as dependent on me, as, in God's way, I think it would have been better for you to have been. I forseaw the straits in which you would be and wished to have you near enough to be reached by that strong right hand as a means of greater love and self-sacrifice on your part. . . . You do not fully comprehend how very set you are in your views and ways of carrying things on. . . .You have been generous in striving to do the will of others, and to put your own judgments in the background. And that very generosity, together with your natural make-up, has caused the nerve strain from which you suffer. Go on doing your best in a large, calm, leisurely way. In due time you will be fitted more snugly into the bandbox of community life.

Take yourself as you are. Bear it ever in mind that human affection is good, if it help your soul to God. The great heart of St. Paul was most humanly affectionate. And St. Chrysostom says, "The heart of Paul was the heart of Christ." God made us to love in Him, and through Him, and unto Him as our last end. It is not cowardly, but necessary for you to do so. There is no self-seeking, so long as your intention is dominantly Christ-seeking. And that it undoubtedly is. . . . A big blessing to you. Be brave and happy.

The nervous strain still affects your feelings: and they in turn react upon your reason. Both feelings and reason are radicated in the same spiritual soul. The nerve vibrations set the emotions a-swinging: an harmonic revulsion of your reason results. Then the reason revolts against its revulsion. For no emotion is long lasting against the reason. Either it is put down by the will, guided by the reason, or it puts blinders on the right reason. Then the blinded reason wrongly dictates to the will.

The remedy: Keep your reason calm. Quietly use your will-power to keep your emotions in check. Be not violent. The Spirit works in calm. Your reason need not tell you that canning beans is the best way to learn Innuit. . . .but your reason, on the authority of God revealing, makes the act of faith that whatever your Superior bids is God's will in your regard. Our Lord bade His followers to obey even the Scribes and Pharisees: "According to their works, do you not: according to their words, do ye." St. Paul urged even slaves: "Obey your masters according to the flesh, not as it were pleasing man, but as it were pleasing God. Serve ye the Lord Christ knowing that of the Lord ye shall receive the reward of inheritance." Your obedience should not be natural, but supernatural. Natural obedience is because of a natural motive, the prudence of your Superior, the wisdom of canning beans, etc. That natural motive has nothing at all to do with the obedience of a Religious. Her obedience must be utterly supernatural, based on one motive, and that supernatural; the faith that what the Superior bids is God's will in your regard *now, now, now.*

Calmly chew on this letter. Get out of it the lesson of obedience. Put your own ideas in N. . . "My just man liveth by faith." Be persuaded that all you do the livelong day is just what God wishes you to do. Let your heart be merry. Think happy things. Tell the devil of discouragement to go to. . . .where he belongs. And let me know what great sport it is to can beans. Stick it out. . . . What might have been might just as readily not have been. So let it go! By slow and dim degrees, little N. . . .will come back to nerve control and then she will be as happy as a meadow lark upon the prairies of Montana. Now she is a Montana bronco. A Montana bronco must be busted. How? By the lash of reason, the lash of will-power, and the shackle of grace and love. However, the breaking in must be gradual: else the high-spirited, self-willed little filly may be made useless. Even the breaking in of a bronco is a scientific affair.

Your difficulties are due to natural causes. They are allowed in sweet Providence for a great good in the spiritual life. You had to be made over, so as to be a good Religious. In your own way you were good, sweet, and pious. But the royal way of the Cross—of self-sacrifice in community, self-abasement in self-rating and mystical crucifixion of your own feelings and ideas—that way you scarcely knew at all. It is one thing to be devout on Tabor and quite another to be loving on Calvary. Self-love shows itself in that you are firm set on your own ideas of perfection. You see the imperfections of others from your own viewpoint. You may be right now and again in your natural judgments from that out-and-out natural point of view. You are wrong in that you do not ignore your own view of what is prudent and take on the view of your Superior with childlike simplicity. Obedience is entirely from the supernatural motive that what your Superior bids you is God's view in your regard. It is a counsel of perfection to obey our Superiors. "According to their works do ye not: according to their words do ye." Religious orders are a concrete expression of that most important counsel of Christ. In the world you find obedience, but not supernatural obedience,—"serving as to the Lord and to man." So send all those natural judgments to Jericho. "Let us go over to Bethlehem." There you will learn "good tidings of great joy." Not the little joys of self-willed N. . . .are at Bethlehem, but the great joy of Christ-love is surely there. . . . There are imperfections in Religion: but they are as nothing by contrast with what goes on in the world. You do not know the world. You have seen it through a wee aperture, and from a narrow, shut-in existence. Oh, I hope you will never go back into it: to face the un-Christly world as I know the thing to be. Take things as they are in Religion.

May God grant you strength of soul and body for the noble, self-sacrificing life, which you have chosen to be yours. Count on Him day after day. He is worth your while, no matter how ought else seem to go wrong.

The following was written to a Sister, shortly before her entrance into Religion, "in answer to a cowardly giving up of the whole thing."

Your turmoil is no surprise; it is the last grasp of the devil at work on your emotions. Your decision was made by reason grace-led, and not by the emotions. Are you strong enough in will-control of those tender emotions, to follow with generosity the grace given? The various reasons, which you detail, are no reasons at all; they are bugbears of the fancy and emotion-bred. The grace of your vocation will be strong enough to meet the difficulties as they occur; so do not trump up barriers, before the time of leaping over them. When the time comes, you will be surprised at your ability as a hurdler. Cheer up! Just take things as they come. You are under no obligation to. . . ; you have made it worth the while to him to train you. Use the organ pedals for all you are worth; your strength will be greater for wielding of the scrub or swab! When things seem blue, just tell the dear Lord: "You are worth my while! A moment of self-sacrifice for You is more to me than a life of self-glory without You."

Hearty congratulations on your birthday—your entrance into the service of Our Lord and the Blessed Mother. This morning, the anniversary of my own entrance into the Society, while thanking the dear Lord for all the graces He has given me, I did not forget the little child just learning to walk along the deep fosse that always leads to Calvary. There is only one way that the mystic spouse of the Crucified can in very decency go, and that is the way of taking up the cross daily, "plunged in grief, yet always full of cheer!" There is no fun along that way, but untold happiness. And be sure that I am praying for your strength and that of N. . . , the strength of Christ-love, and not of self-love, as you both calm your fluttering hearts and wills to "make ever onward to the goal of that so great disgrace," the triumph of the Cross.

Be brave. I know just how you feel. Do not look ahead, unless far ahead into eternity. Take the present by the forelock. Live each day in calm and in love. Thank the dear Lord at the end of each day. Say: "Dear Jesus, that is for Thee: it is not much, but I am happy to offer it to so dear a friend." After the days of "I think I can," will come the long life of "I knew I could."

Hearty Xmas blessings and greetings to you this first Xmas away from home. Again, as often heretofore, you realize, it is no fun to follow your vocation; but a joy past telling. To you is born a Saviour. To you He and His wants mean more than ever in the years ago. There is no sweetness in self-sacrifice; there is sweetness in Christ-love. To give up all, and to be a "cheerful giver" in that giving up, is no easy matter; but to do this for Jesus is easy, at least becomes easy with the years of generosity and love.

How happy your letter has made me! I hope this will reach you tomorrow or Sunday. Often came the thought to answer yours of the Holy Innocents Day; but as often did a prayer take the place of a letter, the Holy Spirit the place of my poor spirit. I knew you would be true; and yet was ever solicitous. Now the first step is taken. How I rejoice with you.

Be sure of my prayers. You are ever with me at that most sacred site, the altar of Jahweh Jesus. I congratulate you from my inmost heart; and bless you with all my priestly power.

Yes, be ever on the lookout for joy,—"Mehr Freude," as the great bishop von Keppler wrote; the joy of the crocus, lifting its bloom above the snows; the joy of the thrush, piping defiance to the March wind; the joy of Jesus, risen from pain and death; the joy of N. . . , mystic member of the mystic Christ; the joy of every soul, that realizes: "to me to live is Christ, to die were gain."

"Look to your call, my brothers;
According to the flesh, not many of you are wise
Not many of you are mighty,
Not many of you are high-born.

God chose what the world counts weak,
To put the wise to shame;
God chose what the world counts weak,
To put the mighty to shame;

God chose what the world counts mean,—
Holds to be nothing worth,
Not to be at all,—
To bring *what is* to naught;
That no flesh should make the boast,
Before the face of God" (I Cor. i, 26-29).

Yes, *Sr. N.* . .will be remembered by me as prayerfully and affectionately, in the mystic union with Christ Jesus, as was little N . . . of years and years ago! When things seem hard, bear it in mind that we are praying for you. Every sacrifice of your self makes for greater grace in your soul, and a fuller conviction that we have not toiled in vain.

No letter from you is in my limbo of the unanswered; but I must send a fatherly Xmastide blessing for great joy, calm, contentment, and love of Jesus.

You are soon to pronounce your vows. Then Xmas will mean to you the swathings of obedience, that mean sacrifice of reason and of will; the poverty of the Babe; and the chastity of His dear Virgin Mother. But to you, still in swadlings, I must send a different message. It is the message of contentment. There is only one way to be thoroughly happy; and it is that of being in sweet harmony with God's ideals and purpose in our life. Without these sacrifices, we should be selfward, not Godward. And to make the road of sacrifice lovable to us, the dear little Babe has first sanctified it. May you love that way of the Babe. May you be thoroughly contented and happy in the rubs and drubs, the ups and downs which are inevitable to one of your temperament.

May all go well with you during these dear days of preparation for your devoted oblation.

Copy of a letter written to a young Sister who had found difficulty with the ways of a new Superior:

No, I am not at all ashamed of you. This attack was due even before this time. Thank God it did not come sooner. You will come out of it a better Religious than ever. The trouble is pathological. Look that up. You are strong in your likes and dislikes; and, as is generally the case in such as you, your nervous system tightens or loosens up, is strained or normal, according to the condition that results from these emotions. There is the diagnosis. Now for the treatment.

First, come back to the principle of obedience. You obey as to Christ. You do not think that what your Superior does, is what the dear Lord would do. You know that what you are told to do, is what Our Lord wishes you to do. Hence, ignore all thought of whether you like the Superior or not, approve of her methods or not. Just think: It is my dear Saviour who wishes me to do this. Always have Him in mind; obey only Him.

Secondly, pray, quietly, and without a fuss; ask the grace to see Our Lord in your Superior, i. e., His bidding in hers.

Thirdly, do not talk of these dislikes, except to those who have a right to know, Superiors, and your confessor.

Fourthly, think of funny things. When down in the mouth, think of Jonas. How happy he was when he came out! That thought may be silly. It is. And, on that account, it will help you to ignore the equally silly attitude of opposition and dislike to your Superiors. Think how jolly pigs are in clover! That thought will do you a heap more good, than will brooding, mooding, in grilling grouch, gnarling growl, and gnawing grief.

Lastly, remember that the devil is trying to make you unhappy. He knows that your final vows approach. He hates to see you in joy.

To another Sister on her entrance into Religion:

Thanks for your letter. May God bless your Novitiate with such blessings as I pray Him to grant you. Two great lessons should result from this year. One is that of supernatural obedience. And to that end, learn that natural judgments have nothing whatever to do with supernatural obedience. There is only one motive that keeps us together in Religion, and that is the principle of authority. All authority, which is legitimate, is from God. Hence we look upon our Superior's will as the will of God in our regard. That does not mean the natural judgment that the Superior is spiritual, saintly, learned, wise, prudent, and what not else. It means merely that God shows us His will by the will of our Superior. That we believe. It is Divinely revealed. The second lesson is that of self-conquest, abnegation, mystic crucifixion: the denial of the feelings, emotions, affections, disaffections, likes and dislikes. That is necessary, if we are to get away from self-love, and to immerse our souls in the abyss of Christ-love, To learn those two lessons is hard. It is a life work. Begin now.

. . . .The following of Our Saviour is a matter of devotedness, of generosity, of good will. "If thou hast the will to be perfect, go sell all thou hast, give to the poor, come follow Me"; oftentimes entrance into Religion depends on this generosity of the will, because grace is not in the feelings; hence the feelings are not conscious of a vocation; the feelings may be conscious of the very opposite. The feelings may be conscious of the sentiment that a good Irish maid in my childhood used to express in a nursery ditty, the melody of which I forget, but part of the words of which are as follows—"I won't be a nun, I won't be a nun, and the Priest of the parish won't make me a nun." The Priest of the parish has nothing to do with it except to direct, and I insist again never become a Religious to please anyone else; never choose an order to please anyone else. Your choosing of an order is your choice, your vocation—not your Confessor's vocation—not your director's vocation; and it would be a very stupid thing on the part of anyone to choose an order because her director chose that order. Very stupid. Choose the order that is your choice, and not the choice of anyone else. The director can throw out leading lines, or life lines, suggest ways, but the final choice of a vocation, a state of life is yours. . . .

Do not worry. Rome was not built in a day! Nor is the temple of the soul. It does not in the least worry me that you are blue, especially after a visit from. . . . That shows what an affectionate child you are. In due time, the love of Christ Jesus will urge you on even more and more. And the sacrifice of your dear ones will not seem to be so dreadful a trial. Then a harder struggle will ensue—in the sacrifice of self-love. But that struggle will not be fraught with the blues. Gradually Christ-love will, in greater measure than ever before, replace self-love in all its forms.

Yes, it is great joy to know that you will soon take your vows. The life will be the same as before. You have lived it most devotedly. So have no fear. The new obligations should not harass you. My only apprehension is that the devil will try to trick you into foolish worry. He knows that you are dear and devoted to our Lord; and sees no possibility of effecting anything, unless it be your unhappiness by worry. If he nags you . . . tell him to go where he belongs.

To a Religious Superior:

Ever and anon come back to the joy of the mystic union with Christ Jesus. Read Colossians iii. You are not so fully hid away as to sign yourself:—Yours dead and buried,—yet the little worries about your soul's perfection are not far removed from the state of the mystic burial of your soul. That mystic burial is the habitual state in which you are hid away. When now and again you seem very much alive to the things that are not of Christ, do not worry; just ignore the thoughts and the worries of them. And even when trying to have continual union with God, let the object of your effort be concrete and "get-at-able." The most feasable union with God to aim at is the very recurring thought of this mystic union in Christ Jesus. He is the Head and we are the members: He is the Groom, the soul in grace is the mystic spouse. So, trying to reach union with God, you will at the same time be dispelling all reason for gloom, and will be cultivating the most patient and motherly attitude towards your Sisters.—And this brings me to what is most important in the life of every Superior: You must be long-suffering, big-hearted, broad-minded, tender, equal and equable to all your Sisters. No hint is meant that you are not such; but every Superior may be more and more such. There must be firmness, but the firmness of a loving mother. That firmness must be tempered by love—and of love, St. Paul writes:—"All things she excuses, all things believes. All things she hopes; all things endures" (1 Cor. xiii, 7).

Note the gradation:—The evidence is at times so clear that there is no *excusing*. Then love *believes* that it must be well. No, the evidence is too strong for such *belief*. Then love *hopes* on for the best. Not in this case; it is hopeless! Then love *endures* it! "Love never fails." Again and again, meditate on that beautiful chapter. As you increase in Christ-love, your Sisters will increase in that Christ-joy which is so important for the well-being of their souls and the efficiency of their toil. Thank God, your Sisters love you and are happy. And yet, you cannot be too Pauline in your motherly love of them.

May your efforts to please Our Lord give you that comfort and peace which is in the mystic Union with Christ Jesus.

CHAPTER XIX

THOUGHTS ON EDUCATION

If there was any subject on which Father Drum thought strongly and spoke with the most unreserved candor it is that of education. He knew at first-hand the teachings of most of the professors of non-Catholic colleges and universities, and took special pains to keep himself well-informed on all religious and educational developments in Europe and America. Knowing, therefore, his temperament and his militant loyalty to all things Catholic, it is not hard to understand the vigor of his denunciation of Catholic parents who would allow their children to be exposed to the serious dangers of a non-Catholic education, for the sake of the social prestige or advancement that was to be gained.

On one occasion a lady wrote to inquire for some good books to recommend to her daughter, who had finished her Freshman year at . . . . College. The child had come home for the summer and alarmed her good parents by the startling views she was expressing on history and religion and other subjects. Father Drum was asked to suggest some books she might read, "so as to get the Catholic viewpoint too." Any one might guess what would be the reaction of Father Drum; but he swallowed his wrath to answer to the effect that there were no such books to be had, none that could, with the mere printed page, offset the influence of a genial professor to whom the girl was listening with rapt attention for four or five hours in the week. To a second letter from the mother, asking what could be done, he replied briefly: "If you want to save her soul, take her away and put her in a Catholic College; if you leave her there, you furnish her with a free ticket to Hell." In point of fact, the girl remained to graduate at this college; and her subsequent ideas of her faith may be guessed from the reply she once gave to a Catholic friend who one Sunday morning invited her to accompany him to Mass. "Oh, I don't need such uplift. I can get more soul-inspiring emotion by communing with my Maker out under the trees."

He was tireless in his efforts to lead a friend and distant relative of his to place her son at a Catholic College. Nor did he cease his urging up to the very end, when the young man was making his last year at the Baptist institution. The sum of his arguments to that mother was that she was giving her son a ninety per cent chance of damnation and a ten per cent chance of salvation. He was genuinely puzzled at the vacuity of the response: "I know N. . . . is a good boy. He has never told me a lie!"

In regard to a similar case that had come to his notice, he considered that a great good had befallen a young woman whose struggle to reconcile the pagan theories of her professors at . . . . College with the principles of her own faith led to a mental derangement that necessitated her departure from College. She lost the degree, but in time regained her reason, and is today a firm Catholic.

In season and out of season he preached Catholic education:

> Give me men with hands that toil, brains that think and hearts that love: you give me perfect men: men perfect in the natural order; men that have fully evolved the powers God gives to man; men ready for higher powers; men whom grace will lift to deeds far and away beyond the toil of hands and the thought of brains and the love of hearts, to deeds of grace on grace and virtue on virtue in this life, and glory on glory hereafter.

Again and again he returns to a plea for recognition of the inestimable value of a true religious education. There were scores of occasions when he went out of his way to bring up the subject, and always he was most unsparing in his condemnation of the irreligious education that has been foisted on this country and generation.

> The state acknowledges its duty to provide for the religious needs of its paupers and its prisoners; but to get state aid for one's religious upbringing, one must become either a pauper or a criminal. The Catholics of this country have for years and years cried to the state the cry of "Bread, bread!" The state has for years and years made answer. "Here are stones!" The bread needed for the upbuilding of Catholics as Catholics is religious education; schools without religious education are stones and fill not the want of the Catholic soul. The Catholic soul needs Christ. Nothing but the grace of Christ can keep out

sin from the soul. No person but Christ can keep a Catholic a Catholic. The public-school system of today has no place for the grace of Christ in its curriculum; it has slammed the door in the face of Christ.

One of Father Drum's speeches on this subject brought down on his head the wrath of a group of anti-Catholic agitators, "probably the forerunners of the Ku-Kluxers," in Baltimore. The occasion and its sequel make up a somewhat curious story.

In June of 1914 he was invited to give the Commencement address at Loyola College, Baltimore. The occurrence was innocent enough, and the address, though earnest and fervent, was no more unusual than a speech of that kind usually is. Part of the speech was quoted the following day in the Baltimore *Evening Sun.*

A few days later, Father Drum was surprised to learn that he had stirred up quite a hornet's nest by his tirade against the public schools. His name was emblazoned on circulars and posters in various parts of the city, denouncing "priest Drum," "the Jesuit foreigner," etc., and a group of those who called themselves "true Americans" wrote to the editor of the *Evening Sun,* anonymously, however, and called upon him to denounce these "Latin dignitaries" who were traitorously assailing the American school-system. The editor was Mr. Henry L. Mencken, the well-known critic and writer. He saw at once the humor of the situation and answered in an early issue of the paper, reserving to the last paragraph of the article a reply that was caustic in its sarcastic brevity:

> That knightly Anglo-Saxon who lately challenged me to embalm in this place a Joint Resolution of Congress aimed at the Papal spy system in the United States now calls upon me (still anonymously) to give space to a long (and anonymous) manifesto denouncing various Latin dignitaries for voicing sniffish views of the American public school system. The only one these malefactors quoted directly and by name is the Rev. Walter Drum, S.J., of Woodstock College. This is what he is alleged to have said:
>
> The education of the will is left out in the courses of the high schools and universities. They are trying to teach the origin of human life, the biology of sex, to children 12, 13, and 14 years old, and they are neglecting to teach these same children the will power which alone can

prevent the child from misusing those things which have been taught.

In addition there is a very brief summary of other remarks by Dr. Drum, in which he is made to argue that this neglect of the education of the will is responsible for "the increase of divorce, murder, and suicide in America."

So far, so good. Examining them most carefully, I can see nothing to inflame the cortex in these asseverations. The learned doctor, indeed, seems to have permitted himself a debauch of platitudes; all that he had said has been said before, and that part of it which was not mere gaudy rhetoric was obvious. Of far more interest and importance than his actual discourse is the comment of the manifesto therein: I quote a few strophes:

"To hear our American public school system vilified and damned by a Jesuit foreigner, and to hear these schools which all true Americans admire pointed at as the cause of murder, suicide, and divorce—and to have them so slandered in the presence of the Mayor of our city, and in the presence of the superintendent and assistant superintendent of schools, is enough to make the blood of every true American boil with indignation. . . . If an enemy declares war against us, and if our public servants, whether civil or military, are found in communication with the enemy, on friendly terms with him, and are even discovered in his intrenchments as a friend of his, what would happen?"

I preserve the English of the original. (Why is it that patriots always know their own language so ill?) Also, I turn to the current edition of "Who's Who in America" to find what godless monarchy of dark Europe spawned this "Jesuit foreigner," this bellicose and plotting "enemy," this damnably hyphenated Dr. Drum. Here is what I find:

Drum, Walter, college prof.:b. at Louisville, Ky., Sept. 21, 1870; s. Capt. John (killed before Santiago).

This episode [wrote Mr. Mencken, in reply to an inquiry], had a curious sequel. In 1919 or thereabouts I printed in my magazine, the *Smart Set*, the first story ever written by F. Scott Fitzgerald, who has since become a well-known novelist. The scene of this story was Woodstock and the principal character was a young scholastic. Father Drum, mistaking the thing for a piece of anti-Catholic propaganda, wrote to me denouncing it violently, and notifying me that he proposed to advise all Catholic women, (whom he said he addressed frequently in their clubs) to boycott it. I replied by recalling the episode aforesaid, whereupon, of course, he saw clearly that I was no anti-Catholic agitator, and a pleasant exchange of letters followed.

The following is part of a speech delivered by Father Drum at the laying of the corner-stone of the new Boston College, on 15 June, 1913, as quoted by the Boston *Pilot:*

Why all this pomp and ceremony today? Is it because of a universal custom whereby important edifices, whether religious or otherwise, are nowadays wont to be begun by the ceremonial laying of the cornerstone? Quite the contrary. We are not following a universal custom. The universal custom follows us. The ceremonial laying of the cornerstone is a part of the Roman ritual of the Catholic Church. Catholic devotion has sowed the seed and an outcrop of this sowing is the now general usage of the laying of the cornerstone of any building with pomp and ceremony.

What is the origin of this Catholic devotion which we have just witnessed? Let us look to Holy Writ. If we look well, we generally find the Church has built her devotions round about the teachings entrusted to her by Jesus Christ through the Apostles. It is the Catholic ritual's expression in concrete form of one of St. Paul's great master ideas—the building idea.

"Ye are God's upbuilding," he wrote to the Corinthians. (1 Cor. iii, 9.) To the Ephesians he amplified this master-idea (Eph. ii, 19-22). "Ye are no more strangers and aliens; ye are fellow citizens of the saints and dwellers in the same house of God, builded upon the foundation of the Apostles and the Prophets, the very corner-stone being Christ Jesus Himself, upon whom all the building being framed together groweth up into an holy temple in the Lord, upon whom ye also are builded into an habitation of God in the Holy Spirit."

According to the Apostle, two buildings are here in question. The first is the holy temple of God which is the Church of Christ, builded upon the Prophets of the Old Law and upon the Apostles of the New, resting upon the corner-stone which is Christ Jesus, growing ever and ever, day by day, in every clime, among every folk, by the setting of stone upon stone, each Christly in form and fashion, taking ever more massive and ever more lovely shape as the temple of the Most High.

The second building in question here and elsewhere in the theology of St. Paul is the soul of each one who believes in Christ as God. "Upon whom ye are builded into an habitation of God in the Spirit." The soul of each of us is either in sin or in grace. If in mortal sin, it is a charnel house of Satan. If in grace, it is a loving temple unto the Most High by the indwelling of the Holy Spirit. And the corner-stone of that living temple can be only Jesus Christ.

The laying of this corner-stone of the magnificent English collegiate Gothic pile, which is the first structure of the group to make up the new Boston College, is only a symbol of the work for which this edifice is erected, the upbuilding of the souls of men into fitting "habitations of God in the Spirit."

The Right Reverend Bishop has prayerfully put the corner-stone in its place; his action, in the Church's ritual, is a symbol of Boston College's putting Jesus Christ in His place in the upbuilding of souls into living temples of God. This is the tremendous purpose of Catholic education,— to put Christ in His place in the souls of men, and in their bodies too.

In the concluding sermon of the Advent course in the Gesu, Philadephia in 1917, after explaining in detail the un-Christian doctrine in vogue at Bryn Mawr, Harvard, Brown, and other Colleges, Father Drum went on to say:

How imprudent then it is if any parent send a boy or girl to such a school! It would be stupidity for our government to send our boys "over there" armed only with 22-caliber Flobert rifles. For over there they will meet with death-dealing 16-inch Krupp guns; they will face a well-aimed storm of ammunition that, blown off at random, destroyed the city of Halifax. No, they must be prepared with armor and arms fit to meet the foe. So, too, our children who go to the great universities for technical and professional training. They will meet no camouflage foe. They will face the most highly organized, the best equipped *Schrecklichkeit*, horribleness, that the devil has ever mobilized for his drives against faith in the Divinity of Jesus Christ. And to send our children against this massed force of ruthlessness with the arms of only a Catholic High-School education is to risk the loss of faith, a loss far greater than the loss of life.

A few scattered selections pertinent to this subject are here given. The letters which follow were written not with the intention of giving a full treatment of the subject of education, but in answer to difficulties of individual teachers who had asked for practical advice:

Education is not the editing of encyclopedias with gilt edges; not the manufacture of walking bureaus of information, more or less accurate. Education is not mechanics. Rather, it is dynamics. It is not carrying dry facts from books to brains. It is fertilization, it is stimulation, it is evolution of the mighty forces of the dormant soul. It is life! Education is life!

"I am come that they may have life, and have it more fully," said the world's greatest educator. What life? All life! Life in the order of nature and life above the order of nature,—life in the super-nature,—cell-life, brain-life, brawn-life, mind-life, will-life, grace-life, life now, life in the hereafter.

It is a great mistake to substitute technical information for education. Once a boy and girl are educated, they are ready for specific and professional information, for technical information. And the more educated they are the vaster will be their range and the better balanced will be their judgment in the acquisition and use of professional information. The purpose of technical schools is to fill the mind up with such information; the purpose of education is to build the mind up.

In Jesuit education, we wish to put the finishing touch to our work; for we are interested in it. An artist would be heartbroken if his painting were taken from him to be finished by another; a sculptor would not wish that the last few strokes of hammer and chisel, the last fine polishing be done by anyone else. Our boys are our artistic product. They stand for our work, and we stand by the impression they make. They are not, of course, the work of one man, but of one moral unit, one body linked by community of aims and means. We wish to finish them off. Our training in philosophy is the last and most important touch we would give them. Is it not fair that we be allowed to do for them all that we started out with a hope to do?

As teachers, we must remember "Jesus began to do and teach." First He did, then He taught. At times we are satisfied to teach. If the teacher shows lack of self-control, she cannot readily teach control of the emotions. It does seem silly to find a pupil trying to steer clear of the jealousies between teachers. "You must choose between so-and-so and me!" What a calamity to the up-bringing of that child!

Written to a discouraged teacher:

Worry not in regard to your class. Calmly do your part, with firmness, encouragement to the youngsters, no scolding, nor scorching, nor scalding! Leave the results to their cooperation, and to Our Lord. It may be that your papers were examined by some young person, too inexperienced rightly to standardize marks. The examination paper may have been inadequate. Lately I saw a set of *their* papers, which must have been made out by one who never taught in high school. There was no gradation. The freshman paper was more difficult than the senior. The papers might just as well been assigned at random to four classes. And yet in no period is the mind of the child so markedly advanced in formation than during the four years of high school.

Your longing for N. . . has to be. You were two-in-one for years; and now nature resents the break. Do not try to undo nature; but build supernature thereon. Good will, and not naughty emotion, counts with an eternal value. Your reward will be exceeding great in heaven.

Do less explaining, and more drilling, grilling, instilling, in the classroom. Take the lower half as your standard. Now and then pique the better students by catching them in error. Look for more from them; and yet do not let on that they are better than the rest. You will get the tricks of the trade gradually. Aim at stimulus—by emulation, pitting one against the other, reading from their exercises to do so. At times a dullard will do better than a bright girl. Read the latter's exercise without comment; then read the dullard's with praise. If you do these things in all seriousness, the children will not be suspicious; the stars will scintillate the more, and the moons will not be utterly eclipsed.

Never worry how to manage your class. Just manage it. Mistakes will happen. Learn by them. Your little ones reverence you, and mean well. They do not analyze you, as you do them. Just think of the days, when you were a school-girl; estimate your girls by the sample that you used to be. There are always a few calculating rascals. They will soon show themselves and you will find that their calculations are not deep. Take them all to be good and true, till proven to be false and naughty. And even then excuse all you may; believe as long as you can: hope, when you cannot believe in a girl; endure with patience and sweetness, when you lose all hope of her. This is St. Paul's lesson to the Corinthians: "All things she excuses, all things believes; all things she hopes, all things endures; love never fails." (1 Cor. xiii, 7.)

In your teaching of music you must look on things in

a big broad way—have a large outlook. Teaching music is inclined to make one narrow and put one in a rut sooner than anything else, because it is all the same thing from morning to night. In the classroom, the teacher is continually changing from one subject to another.

Do not expect too much from the child. Teach quietly and calmly. Do not count vigorously, stamp time or wave your hands wildly to get expression. If you want to say something, do not say it while the child is playing. That will only wear you out, and make no impression on the child. Attract attention by tapping the end of your pencil on the piano, or gently tapping the fingers of the child. Do not talk much. Music is nerve-racking; and the teaching of it is drudgery. Be a quiet teacher.

Aim to give a good foundation, especially in the scales. Make your pupil know her scales, what scales are, and what they are good for. Make her become familiar with the chords of the scales and their relation to each other. Do not insist too much on technic as technic, but try to weave it in pieces, making it as interesting as possible.

Worry not. By teaching, you will learn to teach. Every experience counts. Be reserved, calm, and un-get-at-able, especially at the outset of the year. The girls unwittingly take your measure. And, if you measure beyond their take, it makes for the idea of *tremendousness*. What counts most is that patience begot of love, which you are sure to have.

Theories of education do more harm than good. The best way to learn to teach is to teach, and to get information from your older Sisters. There is very little of true education nowadays; less of it in public schools than in Catholic. The education of the land is under the control of men, who have carved a calf's brain in a laboratory without realizing that a child's brain may differ from that of a calf. They try to make learning easy: whereas it must be hard. Nothing worth while is done with ease. They substitute information for formation. That is their fundamental error. The result is that few children spell decently and write a sentence that is not complete nonsense. Hammer away at spelling, grammar, and clear thought in complete and short sentences. You will in this way turn out prodigies of old-fashioned, formative education. Do not try to ram, cram, jam, slam knowledge into the children. The amount of information they are capable of is too trivial to think of. Another evil of our normal methods is undue training of the imagination. A child's imagination is wild anyhow. Why make it wilder? The result is high-strung, nervous children instead of breezy, boisterous youngsters. You see, I am hopelessly old-fashioned in this. All my education was in Catholic schools, under teachers who never studied the theory of education, and I thank God for deliverance from such tommyrot and twaddle.

With my blessing and the hope that you will cheer up, and teach, teach, teach.

When I taught boys, we used DeHarbe, "Larger Catechism," in high school. Wilmer, "Handbook of Religion" in college. If you look over these, you have what I deem to be the very best eight-year course in religion for girls. The same matter should be taught in grammar, high school, and college; but the manner should vary. In the elementary school, the child is apt to grasp only lead-thoughts; and should merely get by rote the smaller catechism. The high-school child begins to think; but will be more harmed than helped by the solution of difficulties. The college girl will grasp the solution of a few difficulties. very few. Only after a thorough course in scholastic philosophy, is the mind ready, *with safety to faith,* to study religion systematically. In season and out of season the fundamentals should be ding-donged into the minds of the children: only the Church is the rule of Faith: we belong to the Church taught and not to the Church teaching; only the priesthood, hierarchy, and Papacy are the Church teaching. Salvation depends, not on understanding and defending the teachings of the Church, but on childlike faith in those teachings, on the authority of God revealing, and because they are the teachings of the only guide in faith that Christ established. Anything that increases the modern tendency to think for one's self will harm the faith of your children. Catholic Faith is a religion of authority pure and simple; and the only authority that is Divine in matters of Faith, and established for the visible body of the visible Faithful, is the visible Church.

Your points have to do with ecclesiastical history and polity. They are of very little import by the side of religion,—the faith and a knowledge of the Catechism. Do not enter into difficulties: the origin of the episcopacy, the early instances of Papal supreme jurisdiction, etc. Else you do harm to the child's grasp of fundamentals. I am most emphatically opposed to any more than an old-fashioned teaching of the catechism,—graded as above stated, in grammar and high schools. The articles in the Cath. Encycl., e.g. "Pope," "Primacy," "Hierarchy," "Priesthood," "Roman Curia," "Catholic Church," etc. will give you ample matter for your four points. But remember even priests, who are engaged in the ministry, do not bother much about the history of the Roman Curia, etc. Do not aim to teach your children more than the average educated Catholic knows, else you tend to founder rather than found, their faith. For instance, you have "brief history of the internal development of the Church." There is none, except in the modernistic, heretical theory of development. This third point may mean a subdivision of the second: the external development. This external development is by the spread of the Faith among all nations; and by an evolution of the ecclesiastical polity—the jurisdiction of Roman congregations, patriarchs, metropolitans, archbishops, bishops, parish priests, etc. There is no sense in giving that external development, that of ecclesiastical polity, to high-school children. It involves issues of canon law and ecclesiastical history, that only the well-educated, student-priest is able to cope with.

So there you are! You have set me going on an issue that is very much to my heart. I think our schools are suffering from *methodology;* and were better off forty years ago without methodology. Simple Catechism, with side information (from history, Cath. Encyl. articles, etc.)—that is all I would wish our high-school children to be taught! With that information, they may be as strong in their faith; as firm in conviction that Christ founded one Church, and the devil did the rest; as truly and unflinchingly Catholic as is you mother. God bless the Pope.

CHAPTER XX

LAST ILLNESS AND DEATH

Towards the end of May, 1921, Father Drum was invited to address the Graduate Nurses at the closing exercises of the Nurses' Training School of St. Joseph's Hospital, Baltimore. As it happened, the speaker before him on the program, an eminent doctor of the city, made a speech that was a summary of grievances, and proved to be so entirely pessimistic in tone that it left a depressing impression on the audience. Father Drum was the next speaker, and in the nicest possible way began to refute everything that the Doctor had said, and soon had the audience in gales of laughter.

> Yes, shoes are at $15 a pair, but if you work for the glory of God you will get the shoes, at $15 or at any price. And speaking of the indignity of having to scrub floors and wash beds,—why, I remember, many years ago, while I was a Novice, I had to take a month of what we call the "Hospital Experiment" down at the home conducted by the Little Sisters of the Poor.

All this was told in a quiet natural tone,—he was dressed immaculately, white collar, and white cuffs and stately Prince Albert coat:

> I remember that one day an old, old man came in, and you may believe me, he was dirty. Said I: "My good friend how long is it since you have had a bath?" "Fifty years," was the answer, "and the last time I hid wan it give me bronchitis, a cold in me chist,—and I swore thin I'd niver take another." There he was. There was I, and there was the Little Sisters of the Poor; and it was up to me to decide whether that Little Sister or I must take care of him. So I had to go to it. Yes,—but far better would I prefer to scrub floors and wash beds than wash such a patient: and nurses have to do that sometimes . . .. Well, I was talking last year with a graduate nurse, and asked her what the priest had said at the Commencement Exercises. "Oh," she said, "I was tired and did not pay attention to what any of them said. I was through and getting away and that was the end of it."

Then Father Drum turned to the Nurses, and his voice took on a depth of tone that sent tremors through the audience:

> And my dear girls [he said], you are through and getting away; but that is not the end of it. Remember this: though you are leaving St. Joseph's, and will no longer be under the training and care of the good Sisters, and no longer under the strict discipline of a hospital; yet remember, you will be always and everywhere under the watchful eye of an all-seeing God.

He then proceeded to explain the responsibilities of their calling, with a force and seriousness that made his speech long remembered.

Late in the following November, Father Drum was himself a patient at St. Joseph's Hospital. He had really been ailing frequently for the preceding two years with the ailment that finally brought him to death. But he had himself thought, and the doctors diagnosed his trouble as chronic nervous indigestion, and had given him treatment accordingly. In the preceding September, however, a more than usually violent attack of pain sent him to a young physician in New York, one of his former students. This doctor examined him carefully and diagnosed his case as one of appendicitis, and advised an operation. Father Drum had to return to Baltimore first, and while there called on an older doctor, one of his personal friends, who reversed the younger man's decision and pronounced it a stomach disorder for which an operation would not be necessary. It was this indefiniteness about the symptoms and the disagreement among the doctors, that made Father Drum loth to agree even with Dr. Frank J. Kirby, of Baltimore, who about the middle of November urged him that an operation was inevitable, and that it were better to have it attended to then before it became too late. "At least we can wait till the Christmas vacation," said Father Drum. "I shall finish my work and do the usual Christmas preaching, and arrange to be free for the holidays."

Everything seemed to point to a dramatic climax of some kind or other. As often happens under similar circumstances, the events of these few months were bound to take on an almost prophetic significance, though only when all is over do people realize the full import of what is happening. Here is how a nun of the Visitation speaks of the impressions of the last "Uniduum," the one day Retreat which he gave the Community, a week before his illness:

> About nine o'clock the first conference began. His subject was, "The Mystic Christ." As if charmed, we listened, while Father Drum portrayed with wonderful accuracy, the analogy between the Physical Christ and His Church, the Mystic Christ. The three beautiful instructions, at nine, one-thirty and four forty-five respectively, renewed our fervor and gave us food for meditation for many days. Though frequently favored with Father Drum's sermons and instructions, all felt an unusual charm and unction in his words on this occasion. One who had conversed with him at length that day remarked the same indefinable something in his manner. Though as kind and paternal as ever, he seemed more spiritual, more unearthly than on any former occasion. Perhaps he was simply not feeling well; the malady that took him from us may have been already at work. But the impression remains, that the change noted was spiritual rather than physical. When leaving him that day, one of the Sisters said, "Father, we shall see you again before Christmas, shall we not?" He hesitated and replied, "Before Christmas?" She answered, "Yes, Father," then added, "December is a long month, Father." He smiled and said in the same thoughtful tone, "Yes, December is a long month." Those last words have almost a prophetic ring now; for he ended the month in eternity.
>
> The morning when retreat closed, there was more than one tear-dimmed eye. I had been hoping that Father Drum would address us again, though it is not customary to deliver a discourse after Mass during retreat. However, much to my delight, Father Drum having finished the Mass and retired to the sacristy, came out in his surplice and spoke to us again, in a quiet, informal manner.
>
> Those closing words made a profound impression on me. He seemed so deeply recollected. He said that if we had enthusiasm, it would count a great deal toward helping us over the hard places. If we had an enthusiastic love for Our Lord, an enthusiastic devotion to His cause, then nothing would be hard. As an example, he related the Biblical incident of David's three soldiers. When they heard their King was longing for a cup of water, risking their lives, they broke through the ranks of the enemy, secured the cup of water, and brought it back to David.
>
> "Let us have enthusiasm," were his closing words.

"Love leaps over all barriers, where he that loves not, falls and faints. David's soldiers went through the enemy's lines to get a cup of water for their brave captain. Our Lord is our Captain; let us be loyal to him.

As he spoke, I thought: You have that devoted enthusiasm and loyalty yourself, Father; that it is, that helps you over the hard places and keeps you going. I have reason to know that this was, indeed, the secret of Father Drum's success.

The final call to the soldier to halt was soon given. Father Drum was preaching in the Church of St. Ignatius Loyola, Baltimore, on the morning of Thanksgiving Day. Before ascending the pulpit he remarked that he was somewhat unwell and was feeling pain. "The old chronic dyspepsia," he said; —yet he would go ahead and preach the sermon. His theme was the Catholic idea of Thanksgiving, showing how the spirit of the occasion dates back to the days of the Old Testament, and then enumerating the great recent achievements of the Church for which we are to be grateful to God. He alluded to Pope Benedict's recent Allocution on the five great evils of the day, and the Catholic Church's solution for those evils. It was the old theme always, the same battle-cry ringing in this as in every sermon he ever preached, the defense of the Faith against false doctrines, the same clarion: "Remember, stand by the traditional doctrines of the Church!" And although it was noticed that on this occasion the usual profusion of gesture and dramatic declamation was missing—he was evidently suffering great pain all the while—yet many who heard him, declared that they considered this to be the most powerful appeal to loyalty that he had ever made. It was an impressive sermon. His mail a few days later, at the hospital, included a letter from a non-Catholic gentleman who had heard the sermon and wished to go under instruction. But Father Drum's priestly work was then almost over.

Feeling somewhat better that afternoon, he decided to fulfill a previous appointment and called at the home of a friend in town. It was during this visit that he felt a sudden chill, and there was an attack of pain, more severe than any that had yet supervened, which almost prostrated Father Drum by its suddenness and intensity. Yet his one anxiety seemed to be not about his own suffering but about the trouble and embarrassment that he was causing others. He even jested with the physician who was summoned to attend him, and who advised that he be removed to Loyola College and put to bed at once. During the night the pain subsided and he slept quite well; but the next morning, Friday, at about 10 a.m., a spasm yet more severe than that on the preceding day came on, and Dr. F. J. Kirby gave orders to have him removed to the hospital at once.

The operation was performed at 4.30 p.m. by Dr. Kirby, with Dr. A. J. Scheurich as his assistant. Father Joseph A. McEneany, S.J., the rector of Loyola College, was present. It was found that the patient was suffering from acute gangrenous and suppurative appendicitis, which had led to intestinal paresis and sepsis. The appendix had burst and was found bathed in a creamy liquid. It was a type of appendicitis which always gives grave anxiety, because the poison has had time to be absorbed by the surrounding tissues. Paralysis of the intestines often follows, with fatal results. Add to this that Father Drum had in the previous months been taking very little exercise in the open air, and had been engaged in constant hard work at his books, with no respite even in the summer, and we can understand the general weakness and low resisting power that was responsible for the serious complications of his illness. But if the body seemed enfeebled, the will was adamantine. For more than two weeks he held his own. There were days of grave relapse followed by wonderful recovery. As long as there was no certainty that hope was over, the warrior kept fighting grimly and alone against untold odds. Yet he was not alone. All that was humanly possible was being done to aid him. All the Sisters and nurses in the hospital were in solicitude and prayer for him; they knew and reverenced him for his personal relations to them as well as for his reputation. While at Woodstock all were in a fever of anxiety scarcely ever experienced before, and the community was engaged in prayer and in a ceaseless round of watching before the Blessed Sacrament, for his recovery; while Religious communities in convents all over the country were joining in prayer and supplication to Heaven, that his life be spared to the Church and to the Society; while hundreds of devoted friends were watching, in tears and in dread as of some personal calamity that was impending; and while at the hospital, the doctors were applying all the resources of their skill and experience, and calling in for consultation every specialist within reach,—what was the brave soldier himself doing, and what was the nature of the struggle that was going on within the silent walls of the sick-room?

The operation was a most serious one, and although Father Drum seemed to bear up remarkably well under the ordeal, Dr. Kirby had to admit that he was as ill as he possibly could be, and that his condition was very grave. Yet for the first few days there was little fever, and no uncomfortable sensations. Father Drum himself did not deem that his case was more dangerous than the ordinary. With his usual thoughtfulness for others he asked that reassuring messages be sent to relatives and friends, although he was too weak to look at any mail. There were several engagements for sermons and one for a retreat that had to be canceled. He dictated also a few notes to his mother and brothers, in Boston, New York and elsewhere. One was a cheery message to the Woodstock Aid workers who were preparing the big bazaar in St. Francis Xavier's College, N. Y., which he had expected to supervise. An answer soon came back to assure him that all the plans were being carried through, and the affair was in fact a grand success. He asked the attendants to acknowledge for him the gifts of gorgeous yellow chrysanthemums and other flowers that had come during his illness, smiling as he said that someone knew his love for the colors of Sinn Fein. It may be recalled that the Irish struggle was culminating during those days, and that the papers were full of the great proclamation of Irish independence which came in the first week of that December. One of his brothers showed him the big headlines of the proclamation, and all he could do was to smile with visible pleasure and say, "Thank God."

For the last two or three days of November, his condition remained practically unchanged. His pulse and temperature were quite near to normal, but there was great discomfort and pain and he could not sleep well. Still there were no grave apprehensions for the issues of his illness. But on Friday,

December 2, although the patient looked more cheerful, there came a turn for the worse, when a fit of hiccoughing began, that lasted at intervals through this and the following day. Pulse and temperature began to rise, and there was nausea and vomiting. Sunday he was weak and delirious; but almost beyond all expectations he rallied gallantly the next morning, pulse and temperature subsided notably, and he was able to sleep quietly for an hour or two at a time.

But this hopeful change did not last long. For on the morning of Thursday, December 8, the patient sank into a state of coma, and hope was all but gone. The doctors could only say that there was but a fighting chance for him to pull through. It was a losing struggle. All day long he lay in a more or less comatose state, and it was only towards evening that he gradually regained consciousness.

One who was in constant attendance in the sick-room said that all during the ordeal, never once during his illness was there a single moan, or a word of complaint about his sufferings, though it was plain that he was undergoing severe pain. On one of the last days of his illness his brother Joe said to him: "We have named you the Fighting Parson, Walter," and all that he answered, with a sad smile, was "Oh, but it is a hard battle." It is hard to say just when he began to realize that he was going to lose. At any rate, after one of those moments of rally from a spasm of great suffering, he turned to Father Barrett, who had been watching by his side, and asked: "What does this mean, Father? Is it really serious?" When he was given to understand that there was now grave danger, almost immediately came his reply: "All right, then I am ready, Father; please get me my Crucifix."

Father Timothy B. Barrett, S.J., had come in to the hospital with the Rector of Woodstock College, Father P. A. Lutz, S.J., as soon after the operation as was permissible, and had visited him on several occasions during the following days. There was always a priest from Loyola College nearby, either at the hospital or within easy call. But when Father Drum took the turn for the worse on Friday, December 2, and did not improve on the following day, it was decided that Father Barrett should stay at the hospital until the danger was over.

He was due to begin the Triduum of retreat for the Scholastics at Woodstock, on Sunday evening, but one of the Professors was requested to take his place for the opening meditation. The next day, as we have seen, Father Drum again rallied, and was able to sleep for a few hours, and appeared to be out of danger. Accordingly Father Barrett returned to Woodstock to finish the Triduum. Early on Thursday morning, the feast of the Immaculate Conception, he and Father Rector again returned to the hospital, and for the rest of the illness, Father Barrett took a room nearby, so as to be with the sick priest as much as possible.

Shortly after the operation, the chaplain of the hospital, a Redemptorist, came to visit Father Drum, and assured him that someone would be on hand to bring him Holy Communion, as soon as he would be able to receive, every day of his illness. "Thanks, Father," was the reply, "but I am sorry I cannot hear Mass too."

The coma of Thursday left him extremely weak, but he did not again lose consciousness. Early on Friday morning he was seen to go through the motions of saying Mass. He would raise his hands as though holding the Chalice, or spread out his arms in the posture the priest uses at the altar, and move his lips as if he were going through all the prayers of the Mass. He seemed to be in a half dozing state at the time, and on fully awaking, asked to receive Holy Communion. Father Barrett brought him Holy Communion, and all during that day, the patient remained conscious and tranquil, and spoke to each of his brothers and blessed them in turn.

His family had been notified of the danger early in the week, and his four brothers, John D. from Boston, Joseph C. from New York, A. L. from Chicago, and Hugh A. (then Colonel) from Fort Leavenworth, Kansas, all arrived to be with their brother in his illness. Their aged mother was too feeble to attempt the journey from Boston; and their one sister Mary was also too ill to travel, and had to be content to hear daily reports over the wire, and to pray in lonely anguish for the preservation of their loved son and brother.

From the moment of recovery from the comatose state, it seemed as though restlessness had left the sick man. He seemed to know perfectly just what was going on, and there was no flurry or anxiety, and not a trace of fear as to the outcome. It was like going through an ordeal that he had himself rehearsed over and over again. Though very weak, and though his vitality seemed sinking lower and lower, all through the following day, Friday, he was able to recognize people, and spoke to each of his brothers, but could form his words only with difficulty. Towards evening he again became somewhat restless and seemed in great pain. His temperature was hovering in the neighborhood of 103. The end was expected at any moment. All through the night, there were alternate turns of weakness and of apparent rally, though consciousness remained undimmed to the last, and when about six o'clock in the morning of Saturday, December 10, Father Barrett spoke to him and said he was going upstairs to say Mass for his intention, Father Drum looked up and attempted a grateful "Thanks, Father," and could only smile and barely whisper the words. It seems that these were the last he was heard to utter. When Father Barrett returned it was evident that the end was very near. His brothers were summoned, and the Sister Superior of the hospital and several of the nurses and Sisters recited with Father Barrett and Father McEneany, the prayers for the dying. His last conscious act was to press his lips to the crucifix. Death came very quietly. There was no apparent agony. Said one of those who were present: "So peaceful and quiet was his end, that though I never took my eyes off his face, I did not know that he was dead." It was about twenty minutes after ten in the morning when the brave heart ceased to beat, on the Saturday within the octave of the feast of the Immaculate Conception. The struggle, kept up for fully fifteen days, had been hard and unequal; but it was over; and God had not seen fit to hear the thousands of prayers that had been poured out in supplication during all that time, and to give back the silent, austere warrior who had battled against sickness as bravely as he had battled for the Church and for the cause of truth, a warrior to the last. The wires carried the news to Woodstock, and as the Scholastics were filing upstairs from morning class, the great bell overhead boomed the silent, slow knell that, according to Jesuit custom, announces the departure of one of their brothers; and all kneel, wherever they happen to be, to say the *De Profundis* for the soul of the dead.

Father Drum had finished seventeen years in the priesthood; he was just over the age of fifty-one and had lived thirty-one years in the Society of Jesus. He had been building up an international reputation for eloquence in the pulpit, for zeal in the direction of the Spiritual Exercises of St. Ignatius, and for vast acquirements in the field of Oriental and Scriptural learning, a reputation that seemed destined to make him the authority on such subjects for years to come. "Thy ways are not our ways." Once more we are face to face with the great realities. "Thy will be done." May his dear soul rest in peace!

In his last class at Woodstock, almost three weeks before, to the day, he had finished the hour by translating from the language of the Holy Scriptures the twelfth verse of the second chapter of the book of Ruth. It was another of those unintentionally prophetic accidents. The verse reads, from Father Drum's own notes:

> May Jahweh recompense thy deeds, and may thy wages be full from Jahweh, God of Israel, to whom thou art come, and under whose wings thou art fled to seek refuge.

Early in the afternoon of the day of his death, Father McEneany called the four brothers together in consultation, and asked their permission to have the body lie in state in the Church of St. Ignatius, and afterwards to have the Mass of Requiem there. They requested a short time to consider the matter, and finally decided not to have it so, arguing that it was more in accord with Father Drum's own wishes, to be buried as a Jesuit, and at Woodstock. He himself would have asked for nothing else.

Accordingly the body was conveyed to Woodstock, arriving there at about 8.30 in the evening, and was laid out in the parlor of the College. The next day was Sunday, and all day long a procession of friends and neighbors came to pray at the remains. The Mass of Requiem and the Office of the Dead were attended by the entire community on Monday morning at six o'clock, at which the four brothers and a few privileged friends of the deceased were present. But in order to accommodate the numerous friends of Father Drum who had expressed their desire to attend the funeral, actual burial was deferred till the afternoon, after the arrival of the train from Baltimore; and despite a bitterly cold day and a raw piercing wind, they came, to the number of several hundreds, by train and automobile, to witness the simple ceremony of the burial of a Jesuit. The procession started from the door of the College. Every Scholastic was there, walking by twos, then came the Professors, then the choir, and the Rector of the College, accompanied by acolytes; and behind them, the Brothers carrying the coffin. The relatives and friends fell in line behind. Every one carried a candle. The procession wound its way slowly to the little cemetery near the gate; while the choir chanted the *Benedictus*, and at intervals of fifteen seconds the big bell tolled its mournful knell. It was declared to be the largest and most impressive funeral ever witnessed at Woodstock.

"Do you remember the lay of the land near the cemetery at Woodstock?" asked one of the brothers, long months afterwards, "that level space along the cemetery hedge, on the side towards the college? We were passing there one day Walter and I. I shall never forget my awe, when he stopped at that level space, and in a tone of deep sadness and humility said: "That is where I ought to be buried; *near* that group of holy men that sleep around the altar; but not among them!"

To quote from one of Father Drum's retreats:

> Death is one of the punishments of sin, yet we should look forward to it, not with fear, but rather with hope. Our death should be our last act of love, not our last act of fear. Cardinal Wiseman, when dying, was asked how he felt. "Like a school boy going home for the holidays," he replied. Like a school boy going home for the holidays! They who have never gone home for the holidays can never appreciate what the great Cardinal meant.
>
> Oh, well I remember the long dreary trip acoss the New Mexican desert. How all the hardships of living at a distant college, and among boarders who were of Latin-American blood, rather than of Irish-American blood, how all the hard study was forgotten, when I was going

home for the holidays. The wait in the train! And when that old train, way back in 1885, pulled up to the station at Fort Bliss, and I leaped upon the platform, what a joy to me, as my father caught me in a strong embrace! What a joy it was to feel his joy thrilling me through and through! What a joy to see his eyes glisten as in boyish pride I flung aside the lapel of my coat to show him the trophy of the year's hard work!

Like a school boy going home for the holidays! Oh, it will be a joy to me, when I shall be no more that which I now am, to be that which I shall be for all eternity; when I shall meet my Heavenly Father. Now, I shall be ashamed to meet Him, yet glad. I shall be ashamed to meet Jahweh, ashamed to meet Jesus. When my soul is face to face with Him, I shall stand there, with a hang-dog look. But the look of shame will disappear when He bids me,—as I hope and trust that He will bid me: "Come," I hope then He will see the trophies, the trophies of sanctifying grace, and the supernatural virtues rooted in grace . . . and I shall be happy if He only says to me, "Well, you might have been worse. Come, thou blessed of My Father."

The same meditation often ended with the recital of the verses that were found on the desk of Father Shea, a former President of Fordham College, after his death:

RABBONI!

When I am dying
 How glad I shall be
That the lamp of my life
 Shall have burned out for thee;
That sorrow has darkened
 The path that I trod;
That thorns and not roses
 Were strewn o'er my sod;
That anguish of spirit
 So often was mine,
Since anguish of spirit
 So often was thine.
My cherished Rabboni!
 How glad I shall be
To die with the hope
 Of a welcome from Thee!

All the Baltimore papers carried full accounts of the career of Father Drum. The story was told also in the papers of Boston and Philadelphia, and in some of the great metropolitan dailies there appeared a chronicle of the passing of the noted Jesuit scholar. Of all the press accounts, probably the most accurate and concise tribute appeared in the pages of the *Ave Maria* for December 31, 1921:

> The seemingly premature death of Father Walter Drum, S.J.—he was only fifty-one years of age—is a distinct loss to Catholic scholarship in America. One of the most distinguished members of a Religious Order in which distinction is common, Father Drum achieved notable success in many fields of intellectual activity. He was a traveled scholar and a scholarly traveler, a linguist, a theologian, a biblical student, a preacher, and the author of several important books. His "Pioneer Forecasters of Hurricanes" attracted exceptional attention in 1905; and his collaboration with Dr. Sanford in the latter's "Pastoral Medicine" made his name a familiar one to the American clergy. Of late years his exegetical contributions to the *Ecclesiastical Review* have proved one of the special features of that periodical. A many-sided man and an exemplary Religious, he led a strenuous life, which was crowned with a peaceful death.

CHAPTER XXI

CHARACTER AND PERSONAL INFLUENCE

A writer in the issue of *Truth* for April, 1922, in summing up the elements of Shakepeare's greatness, adds a truism that is perhaps as pertinent to our subject as to the Bard of Avon.

> It is surprising to find how little the personal history of a genius really matters when his influence on others is concerned. How and when he lived, and from what stirp he sprung—while interesting and illuminating facts in themselves—are far less important than what he did to deserve the remembrance of men. Suffice it for mankind that he lived, and somehow wrought worthily for its good.

No one will claim, however, that mere external achievements are here meant. A man's worth to the world is measured by the more enduring standard of the good he has done, and by the personal influence that remains in memory to uphold others in the hard struggle of life.

Not the least interesting of the many tasks that fall to the lot of the biographer is that of inquiring for the most striking characteristics that seemed to appeal to one as deserving of record. The replies will be often worthy of study.

A fellow Jesuit and a close friend of Father Drum said, "It is almost impossible to draw the man. There are too many sides to his character." Yes, character is much too elusive to be photographed; much more is this true in the case of a versatile personality like that of Father Drum. The most that can be attempted, and all that has been attempted in these pages, is a picture that will represent him in outline, with just enough of the lights and shadows to throw into relief the more salient features of his character. The following incidents and traits were gleaned from many letters and conversations, and, like a few strokes of the sketcher's pen, will help to visualize the man more fully.

When advised of this biography of Father Drum, his Grace, the Rt. Rev. Michael J. Curley, Archbishop of Baltimore graciously wrote:

> It gives me pleasure to commend this volume. Although Father Drum was not known to me personally, yet his reputation as a Scripture scholar, and a vigorous champion of the faith, has long been familiar.
>
> In many parts of this Archdiocese, under the administration of my predecessor of happy memory, his Eminence James Cardinal Gibbons, as well as elsewhere in the country, Father Drum's voice was often raised in defense of Catholic truth, and to combat error.
>
> In this sketch of his career, the strong soldierly personality of the man, and the zeal and sincerity of the priest have been portrayed with a vigor that ought to appeal strongly to all American Catholics, while the more amiable qualities of Father Drum will please many who perhaps knew but one side of his character. We need more priests like him, men of great moral strength, of deep learning, of unselfish disposition and of zeal that is untiring.

Dr. Frank M. Kirby, a life-long friend of Father Drum and one of the leaders of the medical profession, in Baltimore, said:

> To my mind, the most remarkable thing about him was his loyalty to the Society. He was the most loyal Jesuit I ever came across. There was nothing picayune about the man. It would not do at all to say that his friendliness and geniality and sociability were the more notable characteristics,—they were the least of his many good qualities.

One who was privileged to have Father Drum's special direction thus writes of this same loyalty to the Society.

> Devotion to the Society was so much a part of his very being, it seems impossible to separate it from himself. He lived for the Society, studied for it, toiled for it, loved it with an undivided heart. He would say: "As a Jesuit, I must do this; or I must not do that." "As a member of the Society, such a thing is my duty." Thus in speaking of some sports of his boyhood, he said: "It was all right then; but of course it would not do to run those risks now. For now my life belongs to the Society." And it always pleased him to call attention to the achievements of the Society, or of some individual member of it. For his devotion was wholly unselfish. The Woodstock Aid work he did fully shows this. I recall that he was so touched at the extreme poverty of the Fathers in Austria. he did all he could, in his own quiet way, to collect stipends for them, and soon succeeded in getting together a respectable amount.

It seems safe to say that Father Drum never yet made one friend for himself, but always for the sake of the Society. He had the highest possible regard and esteem for all Religious Orders; but the Society was the mother of his soul.

He loved the inscription that is carved over the door of the little mortuary chapel in the Woodstock cemetery, considering it the most beautiful of all the inscriptions at Woodstock. *Quos genuit Societas*—it seemed so to appeal to his heart.*

Even more beautiful than his devotedness to his family was his intense feeling for his younger brothers in Religion. If one thing more than another drew me to Father Drum, it was the unvarying kindness and solicitude with which he always spoke of the Scholastics.

Once when I was at Woodstock, during the war, the Scholastics' retreat was in progress. I remarked that it was probable no other visitor was on hand but myself. He answered: "I do not know. We are very strict in the matter of not allowing the Scholastics to be disturbed during the Retreat. But last night there was a long distance call over the telephone. The brother of one of the Scholastics is on his way across to the front. He has only a day or two before sailing, and he asked whether he could come to see his brother. Of course, I have no idea what Father Rector did about it. But," and a really pathetic look came into his eyes, as he paused and sighed before concluding, "I hope he let him come. For it is his brother and he may never see him again." I could almost have worshiped Father Drum that.

You could never induce him to stay around and just enjoy himself [said one of Father Drum's brothers]. He generally had but two hours to spare, and he would have to hurry back to Woodstock. He seemed to have a genuine love for Woodstock. Often he would say: "Though I like to be with my friends and do good wherever I can, yet I never feel satisfied and at peace until I am back again in that little room at Woodstock." Really, we called him a "compass" always turning one way, to Woodstock.

How often we are told by spiritual writers that Religious

*The full inscription reads as follows: *Quos genuit Societas/ eorum charos cineres/ coelo reddendos/ sollicite heic fovet.*—"The Society of Jesus, here cherishes, as a keepsake for Heaven, the dear ashes of those she brought forth."

life does not destroy human affection, but rather hallows it and makes it more Christlike and beautiful! Many people discovered in Father Drum the warmth of his affection for his family, that served only to increase their esteem of him, if only because it was so unexpected. This is how one of his friends speaks of this particular trait:

> Surely if there is one thing more than another for which he will be rewarded in Heaven, it will be for his obedience to the Fourth Commandment. His love for his mother was so tender, it was as if she were the very core of his heart. I can still hear the reverence in the tone of his voice whenever he mentioned her. The last time but one that I saw him I asked him if he had seen his mother on his recent trip northward. "Yes," he replied, with a certain hesitating tone, "but it was sad too. For mother is now advanced in years, and at my recent visit to her I have felt, and I know—that some visit soon will be the last." *That visit was the last.* But he thought only of the terrible sorrow it was going to be to him some day to get word that his darling mother was dying, or dead. Never did he dream he would be the one to leave her.
>
> But of no one did Father Drum speak to me so often as of his father. He fairly worshipped that father. His whole countenance would light up as he recalled one or another incident in connection with his father. I have letters in which he speaks of his father most beautifully and affectionately. I remember a scrap-book or portfolio he had at Woodstock. He showed it to me more than once. In it were different pictures of his father and newspaper clippings of his father, especially the accounts of his death and funeral, and articles written by him. One of these articles was on the Anglo-Saxon myth. Among other things there was an envelope, which contained the last letter sent him by his father. How beautiful it was to see how he treasured all these souvenirs and how neatly and carefully he kept them for so many years. I can still see the glow in his eyes as he showed me his father's pictures. He was proud of that father, proud with the honest pride and veneration of a true son's loyal love. I can hardly recall of one conversation in which he did not speak of him, and always in the same worshipful way.
>
> And as he loved his father and mother, so also did he love his sister and brothers. His loyalty to his elder brother was very beautiful. He used to tell me so feelingly of the nobility of his brother's character. He always seemed to look up to him, showing an unfailing res-

pect for his seniority. It was truly edifying, this regard of the younger brother for the elder.

Of his younger brothers he always spoke with the greatest affection. They are a gifted family. He had almost paternal regard for the youngest. During the war, in letter after letter he would mention this brother and so earnestly beg me to pray for him. And after the war, many were the evidences he showed me of his love for this brother. But beautiful, though this filial and brotherly affection be, it is, after all, what one would expect of the ideal son and brother.

A relative of Father Drum had married a non-Catholic wife. She became intersted in the Catholic Faith and wrote to Father Drum to make inquiries, and among other questions, asked how often he would permit her to correspond with him. His reply was as follows:

> Your letters may be as long and as frequent as will help you to be a good Catholic. I have loved N. . . [the husband] since he was little boy, and have followed him affectionately through his boyhood and manhood. For his sake now, and for your sake in due time, there is nothing more important in my life—so far as it is priestly and ministers to others—than in your knowing, understanding and accepting the Faith of our Fathers, the old and only true Faith of Christ.

His loyalty to old friends is thus noticed by a fellow Jesuit, one who knew Father Drum well, and was associated with him from the very first days of the Novitiate:

> The trait that I liked best in him, although it may not be the most striking of them all, was his loyalty to old friendships. You know it often happens in life, in and out of a Religious Order, that in the course of time friendship dies, dies at the roots. Men may be on very friendly terms, but as years go by, and their work makes them separate and drift apart, they come to forget their friends completely; and sometimes never seem to care. Not so with Father Drum. Somehow he had the faculty of being able to keep his friends even after years of separation. He could always be depended on to be loyal. He was of the "once a friend, always a friend" type.
>
> You had to take him as he was. I knew him very well, and often told him that he was constitutionally and naturally "the wooden soldier." He did not mind. It was simply his natural way of acting, though other people

> might not call it natural. His earlier writing had the same quality of the bizarre, the seemingly affected, the intangible something not exactly true to taste that he could not get rid of. He once asked me to show him just where the deficiency came in. Of course it could not be done. In a question of style, of taste, of choosing just the correct nuance of expression and the word that is at once appropriate and in harmony with current usage, one has to feel that touch. It cannot come from without .

Very likely the reference here is to that same "archaic" style of writing that others have noticed. This was nothing more or less than an unusual wide radius of expression, born of his prodigious reading in the best of serious English authors. No one can read a great deal without enriching his vocabulary and widening his command of language. But let a writer once begin to use the powers at his disposal, and he leaves himself, naturally, open to the criticism of being unusual and bizarre.

The influence Father Drum exerted over Religious and their reverence for his character is sufficiently revealed elsewhere in his letters and retreat work. A Religious who had the privilege of many years of friendship with him wrote this estimate:

> He seemed to have a genius for friendship. He gave himself, his time and his talents completely to his friends. Two faults only he never overlooked: indeed, they incurred the forfeiture of his friendship: these were insincerity and the failure to respect his priestly character. With all his charm of manner, his innocent gaiety and sparkling wit, there was a certain dignity, one might say majesty in his bearing, that ever reminded one of his sacred calling.

According to one estimate, really inadequate and misleading, his was an unusual type, an extraordinary personality that, combined with a great deal of learning and a certain amount of bluster and bravado, has an irresistible fascination for a certain type of imagination, particularly feminine. He knew his assets and played them up to the limit. If this be granted, though it is by no means a truthful or an adequate estimate of his influence,—call him extraordinary, or picturesque, or bizarre, if you will. Even then it is impossible to understand why some people wish to quarrel with a character that was come by naturally, and used honestly and nobly for a sacred purpose. His use of his natural gifts was for no selfish purpose, for no unworthy end, but always and only for souls to save, and to gain for Christ.

But it is evident after a little reflection that the influence of this man was based on something more solid and durable than mere personal charm. In the first place it will be enough to consider how numerous and varied were the characters for whom Father Drum's influence was a constant force, leading them for years to devoted loyalty. There were very few who once came under his influence and afterwards gave up his friendship. As many as eighty-five young women were finally guided by him to Religious life, after being for a long time under his spiritual direction; ten young men became Jesuits, while with numbers of other people he kept up a sympathetic correspondence that helped them in stress of sorrow as if he were the only guide and father to whom they could turn.

Besides this, it is evident that an influence which lasted for years and was sometimes characterized by a stern and inflexible rigidness of purpose, and a forceful direction that refused to bend to the whims or mistaken inclinations of those whom he directed, must have been based on a knowledge of human nature and a power of moving others that was a permanent source of strength. One striking characteristic that stands out in his letters of direction is his ability to concentrate his attention, and unify his efforts for a definite end to be gained in the training of each individual soul, without seeming even to suggest to the subject that he or she was but one of several hundreds other to whom he stood in the relation of spiritual father and guide. It was as if he were able to individualize himself with each one as the sole personal friend.

Naturally, such an influence does not often admit of being affirmed by verbal testimony. But there are many willing to come forward and bless him, for the good he had done to their souls!

> It was in 1907 that Father Drum began to exercise on me an influence which had greatly helped in the shaping of my life. I cannot well describe the power he had over my heart and mind. In my childhood he saturated me with the idea that the service of our Lord consists in a loving dependence on His mercy, and not in the slavish exactions of fear. "Almighty God is not a policeman with a big stick," was his favorite argument against scruples.

The great man's personal holiness and his vast knowledge of the sacred sciences, together with his tactful sympathy made me turn to him for the solution of many of the problems of life. He insisted that the only kind of spirituality is based on right reason. The word "reason" and "Father Drum" are almost synonyms for me.

For years, his very name has been like a battle-cry urging me onwards in the service of the King. His intense interest in my spiritual welfare, and his deep affection united him to me with "hoops of steel." I have often been impressed by his reverence for our Religious customs. This characteristic is shown by the following fact. Despite our close relationship—he was my cousin—he always addressed me in conversation by my Religious name, from the day that I received the habit. That I belong to a Congregation approved by the Church, was a real happiness for him.

Father Drum's dealing with souls was so gentle that it required no great act of faith to see Christ in his person, in the confessional; it was so strong that it was impossible not to feel with certainty that he saw only the soul. He seemed to have but one object, to point the way to God. Except for this purpose, Father Drum never said a word or withheld one. For though he was gentle, even tender, he did say strong things, words that wounded with a wound to heal. An unexpected event had so disturbed a young woman who was on the point of entering Religion as to make her bitter and unresigned, though she did not waver in her purpose. "You are rash to court failure! A clinging vine! A leaning tower! Where will you find generosity to persevere? You will be a gray, discontented half-baked individual!" He wrote long afterwards. . . "The results have fully justified my treatment, to which you responded as you would not have done to a milder one." Naturally speaking, Father Drum's insight into character was uncanny. Supernaturally, it can be explained by the fact that he lived and worked by the light of the Holy Ghost. Once, he gave in the second of two short interviews a character sketch in two phrases. It was not intended to flatter. Five years later, new circumstances brought out the latent tendencies, once revealed in those two phrases.

Father Drum's regret when he could not give assistance was deeper than one's own disappointment at the loss of his help. In a heavy family trial which one member had to bear alone, Father Drum's influence would have removed a load of sorrow and anxiety. His distress and what was almost indignation at not having been appealed to, no matter what the distance was, were proof enough that nothing short of the impossible would hinder him if there were question of a service. "If I had only known," he said over and over again. In trouble one felt in him the strength of a support that was like God's made visible, a support that imparted its quality of stability. "The downs of life are formative of Christ in you," is typical of the pithy, vigorous style of his response to one confiding to him a trial. A natural diffidence of character and a positive shrinking from giving confidence had resulted with one soul in a state of spiritual and mental uncertainty and doubt which was well-nigh unbearable. This person had just graduated from a non-Catholic College and had had many difficulties in matters of faith, which were consequent upon her education and her lack of knowledge of Catholic philosophy. For the first time in her life, she felt free to express her difficulties and aspirations to Father Drum whose insight was so clear, that his treatment and direction entirely changed the course of her life. His first letter was strong and almost sarcastic, if Father Drum could have been termed so at any time. It was followed on her part by an angry silence for two weeks, when grace triumphed and she answered humbly and gratefully. After, that, his treatment was less severe and he said many years later that his words at that time were more severe than his judgment had been, but he realized that the spur was needed. Usually, as was the case in this instance, only one of Father Drum's spurs was needed for a lifetime. The words one can never forget, "Your mind has a Protestant bias, due to the education fond parents have unwittingly given you."

Father Drum's fearlessness in face of opponents, his uncompromising loyalty to the Church he loved, was alone an inspiration. Often, I have heard him say that one *must* be narrow with regard to principles, but broad with respect to persons. I was once accused of narrow-mindedness by a rather compromising Catholic person of my acquaintance; Father Drum wrote me; "Yes, you are narrow-minded; but with the narrow-mindedness of Jesus Christ, in defense of truth. God bless you."

Many men, many minds. One of the perennial sources of surprise in life is the variety of judgments that may be passed about the same individual. This ought to be expected, of course, when the character of the individual is almost kaleidoscopic in the multiplicity of its traits. Another Jesuit thus sees Father Drum:

The trait that most vividly stamps his picture in my memory is his orthodoxy. He was like a watch-dog straining at his chain. If not alone a heretic stalked in view, but even if a suspicious word or gesture came into his notice, he was at once aroused. This is the only way to understand his antagonism of N. and N. His hatred for any interpretation that conceded a compromising point to the rationalists amounted to a passion. The last confession that he made to the class, I recall, was his heartache at N.'s rough-shod letter, (worded diplomatically enough) about the disagreeableness of over-orthodoxy.

To me, the next important characteristic in him was his *student-ways*. It will be hard to explain this without suggesting odious comparisons that are unintentional. But the plain, unvarnished truth is that he was the most devoted man of books that I knew in the Province. In my mind this amounted to almost a fault, and perhaps that is the reason why he never produced a lasting book. He was so well-read, so penetrating, so sensitive to the varied viewpoints of a mooted question that he quailed before the task of putting to writing the half of what he knew.

Lastly, but not least, was his fidelity to Religious duties. I mention early morning visit, and his devotion in saying Mass. With his knowledge of Scripture, his unquenchable enthusiasm for Paul and John, the Mass in all its parts was actually declaimed by him for the great love of Scriptural thought that filled his mind.

My personal relations with him were pleasant: especially when on journeys. He was a charming companion, and his conversation usually took a serious tone, in which was absent the *ex-cathedra* of the class. He invited discussion and was somehow able to hide his own knowledge pleasantly. . . As you know, I always substracted the pompous and dictatorial in him to enjoy his white-hot orthodoxy. The more I think of it, the more I am convinced that the inspiration of his noble traits, and the reason of his "puzzling excesses" was orthodoxy and nothing else.

That Father Drum himself lived the loyalty and faith that he preached to others so incessantly was patent to all who knew him; it was a visible fact. In his daily life he was simple, earnest, mortified, tireless, and never wasted a moment. He seldom attended even the special lectures and entertainments which on rare occasions might be provided for the members of the community. His was a scrupulous regard for all the detailed regulations of Religious life, making it plain, at least to the exterior eye, that he was ever faithful to the sacred vow that he had taken in the Tertianship, never to violate a rule deliberately. His edifying devotion to the Blessed Sacrament and to the Blessed Mother was evidenced at Mass and in his frequent visits to the chapel; with no pretentiousness in the matter of dress or in personal appearance, unless one wished to mention the apparent affectation of often allowing his hair to grow rather long. For this, however, he himself gave the simple explanation that he disliked to frequent the public barber shops, and he, therefore, had to put off as long as possible the necessity of having some one of the community as is customary at Woodstock, be his barber. Urbane and courteous to everyone who addressed him, and always on the alert, seemingly, for the opportunity to be of service to others, there were good grounds for the general opinion that in his own life Father Drum was the hard-working student, the zealous Religious and the blameless priest of God. A Scholastic wrote from Woodstock:

> Those of us who heard the continual tapping of his typewriter in the silent hours of the night; and who saw him in the first hours of the morning kneeling before the Tabernacle to receive a blessing upon his new day; and then at night coming to pay the daily farewell to his Love, know what a student Father Drum was, and recognize his sanctity and simplicity of heart.

Indeed we wonder if this artless tribute from a friend's letter needs any qualification; "As a gentleman, Father Drum was, I believe, the very embodiment of all that makes for the good, the true, the beautiful; as a priest, an *alter Christus,* with all the zeal, love and earnestness of St. Paul."

In fact no one could fail to note the sincerity of his dealings with others. To quote another tribute to Father Drum:

> At times, in his talks with a penitent, one caught a glimpse of his interior life, which filled one with an almost reverential awe. On such occasions, it was easy to understand the fascination, the personal charm, our Divine Lord exercised over all with whom He came in contact. . . . The realization that his words of counsel were the real expression of his own daily life, was too evident to be missed. These words he often said of Priests: "We must

be other Christs." I never left his company without the conviction that he acted as Our Lord might have done in the given circumstances. Indeed personal love of Christ especially in the Passion, seemed to dominate his every thought. "It is not humiliation that we seek, but Jesus Christ humbled and despised," was one of his frequent sayings. And another: "I rather hate poverty, but I have learned to love Jesus poor and despised a little."

His own words unconsciously reveal the quality of this faith that he lived up to:

> Our attitude on faith is that of a litle child, who takes his mother's word for everything. It is only by slow and dim degrees that the child at last ceases to be a child, leaves the wonderland of childhood and realizes that there are some things the mother knows not. The loyal and logical Catholic never ceases to be a child in his attitude towards Holy Mother the Church. The more he studies, the firmer is his conviction that she knows all there is to know in matters of faith. He never leaves the wonderland of childhood in faith: faith's nursery is ever his great joy. "Unless ye become as little children, ye shall not enter the Kingdom of Heaven."

Following the thought of the world-famed Pasteur, who after years and years of study felt happy to have but the simple faith of a Breton peasant, Father Drum makes in a St. Patrick's sermon, this boast of his own Irish heritage:

> After a whole lifetime of study; after years and years given to the original languages of the Old and the New Testament in the early ages, I can make to you, on the eve of St. Patrick's day my simple candid boast that mine is the faith of my Irish father, who was forced to come to this land of freedom by the tyranny of an alien people; and before my death, if God gives me years and years more of study, my fondest hope is that mine will be the still simpler faith of my Irish mother, who was obliged to come to the United States, to live and let live, because of the famine of '48, wrought upon St. Patrick's land by the tyranny of an alien people. And in regard to that tyranny I feel no resentment whatsoever. My attitude is that of the Catholic Church towards the sin of Adam. As the Church in her liturgy praises God by the words: "Oh, blessed sin that has brought to earth so great redemption." I make my simple chant of praise of the tyranny of the alien people that has brought to the United States so great a glory as is shown by our Catholic hierarchy, priesthood,

Sisterhood, and laity. God grant that you, descendants of those thus forced to the United States, may never be brought by wealth and class and self-sufficiency to give up that precious heritage which St. Patrick gave to you.

That he was withal ever on his guard against that pride which is the especial foe of men of great natural talents, is clear both from his unswerving obedience to the least commands of Superiors, often under circumstances of peculiar difficulty, and from the humility of his frequent requests for the prayers of others, in his letters and especially during time of retreats. "The devil, too, is a fisherman," he would often say; "he too will fish for game fish. How he must fish and fish to capture priests! Pray for them." Once during his last illness, one of the doctors in attendance remarked to him jestingly that this sickness was sent to him as a warning against elation of spirit, after the wonderful sermons he was preaching. Father Drum thought a moment, and then said seriously: "No, Doctor, I think I can truthfully say that I never think of my sermons in that way. I have never taken any pride in them. It is simply a part of the day's work, a means of doing some good."

The following is the lengthy appraisal of Father Drum's character by one who knew him well:

Father Drum was a man of whom it seems impossible that word should come that he was dead. So utterly alive, so sensitive to all of the finest things of life, so full of strength and power and the militant soul of Christianity, how could one think of that ardent tender spirit in the quietness and repose of death?

When he came into a room, with that air of perfect gentility, of sustaining, inward happiness, the room might be poor and squalid, or stiff with the formality of luxury, yet he came as though the whole world was his Father's house, and he was perfectly at home and at ease in any part of it. His very personality was like the opening of a window in an overheated room to let in the fresh, cool wind of heaven. One never knew just what he was bringing with him; he was, as it were, always hiding in his hands some hidden gift, as though he said, with a boyish delight, "What do you think I have for you?" Perhaps it would be only a little song at which he had laughed and which he would sing to amuse a child, sing with the simplicity and understanding of their strange wistfulness, their bashful advances, that made him their adored companion. For not all the dust and weariness of the world's labor could ever wither the flowers of childish enthusiasm he bore in his heart, and the children, with their unerring instincts, recognized their kinship with him, and made him at once one of their joyous fraternity. Perhaps it would be a treasure from his many-sided intellect, a translation of a Latin poem, begun, with an exquisite feeling for the pathos of "old unhappy, far-off things," a beginning to tantalize one with an unappeased desire for a lovely completeness; a completeness not to be realized because of the deeper interest in his Oriental studies that finally absorbed, but never made him quite forget, his early love of the classics. How often one can remember hearing him quote certain passages of Homer, the magic quality of his voice lending a deeper significance to the sonorous Greek. Perhaps, it would be some curious philological analogy from the dozen or more languages he had always at his tongue's end, some fine translation of a Psalm making its poetry more beautiful or giving new and wider meaning to its thought, yet ever within the limits of the understanding of his audience and all given without a trace of pedantry or self-consciousness. Always there was something in his hands, something real and unforgettable, to be given with joy according to the needs of the listener. But best of all he loved to give the thing that was himself, his inner life, his *raison d'être,* which was the love of God. This indeed was his life, this was the cause for which his militant spirit fought and toiled, this was his thought and care, his comfort and consolation, and his exceeding great reward. He could no more help trying to give voice to his zeal for the glory of God than a rose can help its fragrance, and he never desired to do so. To set straying feet in the right way, to put idle hands to work, to fill empty heads and hearts with something worthy of their maker, this was his constant labor, done with a gentleness and kindliness born of that sensitiveness to pain that made him so careful of ever hurting anyone else. He had an unfaltering sense of justice, yet the tenderness of his heart was such that he succeeded where many another failed utterly.

One thinks of him, in the early days of his priesthood, toiling amongst the poor and the sick on Blackwell's Island, a young soldier on his first fields of conflict, working with an austere devotion to duty only possible to youth and utter vocational dedication. One sees the trend of his well-balanced mind, and the great wells of his affectionate nature, wells whose first gifts of pure water must always be for God, and afterwards for the needs of his fellow men, in the little commentaries written only for his own eyes on the various books he read during his vacations at Keyser Island and at Woodstock. With what naive and youthfully solemn judgment, with what sympathy yet unerring sense of rectitude he measures the lives of the characters, summing up in a few words the probable effect of the books on young and untried minds, and expressing his own pleasure in all that was really fine, and his condemnation of anything that was small and mean. With hypocrisy and injustice he had no patience, neither any compassion on the falsifiers of history, but the noblest and most touching things of art and poetry made an instant appeal to him.

How many ways there are for one to think of him, how many little kindnesses, how much of tender thoughtfulness, unwritten, unrecorded, save in the hearts whereon they are indelibly impressed.

No one who heard him preach could ever forget the experience; there was about him that controlled, burning eloquence, as he strove to share with everyone the vital things of the spirit, an eloquence so powerful that the very words were lost in the great vision he revealed in an exquisitely sad resonance of voice that still falls upon the inward ear in immortal music. The Vision itself was all his care; he had very little thought for any of the efforts of his mind that had made it, yet those to whom he revealed it saw and remembered and understood. One thinks of him with almost an awe in the midst of his profound studies whose extent and magnitude none save a scholar could possibly comprehend. One sees the vast outlines of such exhaustive research as one looks at the huge bulk of far-off mountains but the great dark precipices, the winding roads, the towering trees and the lovely expanses of flower-starred sward, are hidden by the mysterious blue distance, and like a traveler's tales come the glimpses of those distant places, seen dimly through the medium of another's intellect. Only the traveler himself knew all the toil, the monotony, the unsatisfied yearnings, of those strange roads, the glory, the adventure and the silence. And all the cares and responsibilities that the years brought him, could never quench the burning ardor of his work; none of the burdens that he bore so manfully and with such pathetic patience could dim the brightness of his childlike gaiety; always one thinks of him laughing along the difficult road of life stretching out his hands in his own extremity to help his fellow travelers.

And still he laughs along the way with us, for in

thousands of young minds the strength of his powerful intellect urges on to higher ideals, nobler aspirations and stronger faith; in thousands of souls there are white places of purity cleansed by contact with his own pure soul, and in more hearts than it often falls to the province of one heart to influence there are joys of loving recollection seen clearer through the tears of warm, human affection.

And he himself has joined that great company of the invisible hosts of his fellow warriors of God, who strove and fought in toil and persecution, whose strength spurred on the weak and upheld the strong, whose voices rose like music from the desolation of the wilderness and the blood turmoil of the seething cities, whose souls and bodies were a living oblation for the glory of Christ and His Kingdom, lifted to God like the incense of an evening sacrifice from a world that could not understand them nor content them. They are not dead, but their burdens are lifted, their hands are free, they see the beginning and the end of their labors and understand the broken dreams and griefs of life, for though, in the sight of the unwise they seem to die, they are at peace, and God Himself has satisfied them.

> "Gracious God keep him,
> And God grant to me
> By miracle to see
> That unforgettably, most gracious friend
> In the never-ending end."

Of course, in estimating a man's character, nothing else matters but his motives. It is not what we do, that counts at all in our efforts at perfection, natural or supernatural. It is why, and how we do it. Many make the mistake, as we often see when Lent or a mission comes along, of giving up this or that trifling material thing, retrenching here or there in non-essentials, when they should really be looking to their motives, and purifying their intention. Is this being done for some definite good, or am I simply doing it for pleasure's sake, or to pass the time, or am I aiming at some unlawful purpose?

It was conspicuous in Father Drum that he seemed never to be acting on mere impulse or without a definite, ulterior aim; it was the case of a man who was a rigidist as far as he himself was concerned, in guiding his conduct by the main principles of Religious life. Does this help—in a definite way? Is there any good to be had for some one? He asked once in a retreat conference:

How am I going to live? I wrote very extensively along these lines in my Noviciate days. I tore up all those notes long ago. Take your rule book. That is enough.

That he had early schooled himself to regulate his conduct by strict motives and right principles, rather than pleasure or impulse, is seen most clearly in the diary of his travels, where there is abundant evidence of his fidelity to daily Mass, to the requirements of Religious poverty, and to those Religious practices that are so apt to suffer neglect amid the distractions of strange scenes and the absence of example from others. One little incident may serve to illustrate the nature of his conduct under rather distracting circumstances. A friend sends this reminiscence:

It was about 1909 that we once induced Father Drum, after much urging, to come with us to visit N. . . for a needed rest. The visit lasted only for a day. About noontime we were strolling on the gaily thronged boardwalk, Father Drum, my brother and myself. On arriving at one of the pavilions he suddenly turned to us and asked us to wait for him there as he had to leave us and say some prayers. In a little while he returned. The matter had no special significance for me then; but since I have entered Religion, I know that it was the particular Examen that he had taken pains to make at the prescribed time.

A few anecdotes occur to mind that might find their place here to throw on the character of the man a light by which he is perhaps little known. Many people in one of our parishes in a large city remember the occasion when Father Drum, coming out of the Church after a lecture that had created tremendous enthusiasm, saw a crowd gathered down the street. Pushing his way in quietly, he found it was only a poor fellow lying on the pavement, maudlin drunk. Instead of turning away, he bent down and picked the man up, got his name and address, and then started to help him down seven blocks to his home. He had almost to carry him, and then led him up four flights of stairs to his shamed wife and children. The next day, Father Drum actually called again to see how things were, and helped the man to find work. Next followed his conversion—the man had been away from the Sacraments for years—and he is now living a good Catholic life. It is hard for many who knew Father Drum, with his dignity and punctiliousness and all that, to associate him with such an act.

In all the bustle and hurry of his preparations for going abroad, in August of the year 1906, he heard of a young man who was anxious to enter the Society of Jesus, but was hindered by external difficulties which could only be settled by some one on the spot, in his home on the other side of the ocean. Though unknown to him, Father Drum took the time and trouble to hunt the young man up, found out all about the difficulties and promised his full assistance. Six months later, he wrote to explain how all the difficulties were settled, and that everything had been attended to. The young man was able to enter the Society.

Father Drum chuckled when he received the message of the elevator man whom he met in his visits to his brother's office in New York. Of course, Father Drum made the best of a chance to urge the man to correct the carelessness of his ways. One may judge of the nature of the counsel from the message given to Mr. Drum! "Yes, sah, give Father Drum my love, but never my address."

After one of the Lenten conferences in the Church of St. Paul, in Baltimore, a wealthy lady offered Father Drum the use of her automobile to take him home to Woodstock. The ride is but seventeen miles. Father Drum thanked her courteously, but declined because, he said, "his own car" was coming for him. In a few moments Woodstock's auto-truck drove up to the curb, and with a pleasant laugh he said, "There's my car," and bade adieu to his friends. But he paid dearly for the edification that was given. It was mid-winter, and that happened to be the coldest night of the year. Before they were halfway home some accident to the engine made it necessary to stop, and the Scholastic who was driving had to work more than an hour to get it going. He did not succeed. The engine froze in the delay. Then both had to trudge to the nearest house, two or three miles away, and get in touch with a garage to send out help. It was nearly midnight when they reached home, frozen and exhausted. How they both escaped pneumonia is a mystery. The Scholastic was laid up for two days. But Father Drum was up the next morning at five, said his Mass as usual and conducted classes that day.

It was a common experience for him to go off of an afternoon to Washington or Philadelphia or even New York, preach or give a lecture and return to Woodstock for the next day's classes; and invariably, on such occasions, one would see him striding down the refectory towards the kitchen, to get the lunch that he always carried with him on journeys. It would be just a sandwich or two. He seldom or never went to hotels or restaurants, unless invited by another, and even on the longest journeys could not be induced to take a Pullman car for greater convenience. This was nearly his undoing on the journey to give the retreat at Notre Dame in Indiana.

One day the pastor of the Church where he had preached, helping him on with his overcoat, remarked on its fine quality. "Yes," he said jestingly, "you Jesuits take the vow of poverty, but we keep it." Father Drum replied; "I have had that overcoat for thirty years; and besides, it was a gift to me when I entered the Society."

It was often remarked that he was very faithful in what others might consider trifles, like asking permissions. One day he walked into Father Minister's room, his hands behind his back. "Can you guess what I have today?" "No, what is it now,—a gold watch?" "Wrong, guess again." "A roll of bills?" "Wrong again! I have a roll of *silk stockings!* a half-dozen pair. Think of it! The audacity of the dame who dared to send me *silk stockings!*" But after a pause he said, "But I am going to surprise you, Father, by asking you to let me keep them. And I'll tell you why. Two weeks from now I am having another bazaar in New York, for the Woodstock Aid. They will come in handy for sale at the bazaar."

> Now I know what it all means [he said in a retreat]. I never felt any difficulty about obedience, or about asking these trivial permissions, until the last few years. The men whom I have to ask for many of them are men whom I taught. The Ministers, generally, are young men, all of whom except one or two, are younger than I; so that if I go to such a man and ask for a permission that a man of the world would consider trivial or absurd, it makes me conscious that obedience pinches. The mortification of the will brought about by asking permissions has its effects upon the soul. It is a form of salutary interior penance.

The writer remembers coming to see him one morning,

shortly after he had arrived home after the Christmas holidays. The bed and floor about it were piled high with various sized packages that had come in his absence. Two hours later, I had occasion to call on him again. There was nothing left. "Where are all your presents, Father Drum?" He showed me a slip of paper, on which he had noted the names of the senders for acknowledgment. "They are all in Father Minister's room. Go and help yourself."

Once before his Christology lecture in the Nardin Academy, in Buffalo, he was standing in the hall near the main entrance, chatting with a group of friends, among whom was the Principal of the Academy. He noticed two Scholastics entering the hall for the lecture, and immediately bowing to his group and excusing himself, he came over to meet the Scholastics and escort them to his friends, to whom he made the introductions. This act of courtesy was only one instance of many similar ones that have been related of him. He invariably made much of any Jesuit whom he met anywhere.

Most of his other characteristics must be left to appear from the course of the story. It really seemed true that in his genuine zeal for God's work among men, in his desire to win for God every one with whom he came in contact, from his earliest years he considered it a duty not to neglect a single one of his many talents. He was, indeed, by nature richly endowed. Some have said they knew Father Drum only as the austere Religious, the man of immense erudition, but no more. It is easy to understand that view, when we consider his habitual reserve and the deep sincerity that he felt in himself and demanded of others. But they had only to make the approach, to find that here also was the delightful, cultured conversationalist, at home on the most varied of subjects, and ready with a seemingly inexhaustible store of anecdotes of travels and experiences, and scraps of poetry; the inimitable impersonator of character and dialect; the singer of droll negro melodies, and possessed of musical knowledge of no mean order. There are hosts of admirers to whom Father Drum was known and deeply loved for just these characteristics. And he was decidedly the friend of children. This someone has called the surest mark of sanctity. Children were attracted to him wherever he went, and we are told that his retreats to the children and to students were actually the most successful he gave.

Finally, whatever be one's individual opinion of the character and personality of Father Drum, there are literally thousands, if we may credit a multitude of letters and reports, who would not hesitate to subscribe to this estimate, in Shakespeare's language:

> His life was gentle, and the elements
> So mixed in him that nature might stand forth
> And say to all the world: This was a Man!

Yet it would be far from just to estimate his work or his worth from mere personal influence, or from any other single characteristic. We must take his life as a whole. That one half hour with the Holy Father lifted Father Drum to the level on which alone he was thenceforth to live—and be judged thereby. "Remember, stand by the traditional doctrines of the Church!" To fulfil that mission, all his studies in sacred and profane literature; all his researches in Biblical and modern languages, all his delving deep down into exegetical and theological and Patristic learning, his class-work and lectures and retreats and preaching and writing (scattered and fragmentary though these last may appear)—all represent an achievement of really large proportion, that was being wrought into one great unity. Father Drum had not yet published any books. But if we may conjecture what that ultimate purpose of his life was to be, from the one thought that permeates and constantly reappears in every portion of his writing and preaching, we may say that all these elements furnished the material for the "domes supporting domes" of the great Byzantine cathedral that was to be his upbuilding of a life-time; and the dome that would crown the whole structure was to be one work, representing the one idea that dominated his life—to spread among men the solid, scientific proof, and the convincing exposition of the Divinity of Christ. This seems to have been the dominant thought of his life, as the words of Pius X were his battle-cry and his inspiration. The last words of every retreat invariably echoed this purpose. The *fervorino* at the end usually closed with this thought: "I could hear the cry of Ignatius of Antioch as he cheered on his followers, the followers of Christ, with the cry of St. Paul: *Jesus Christ, yesterday, and today, the same Jesus Christ forever!*" And in the same strain, the most famous of Father Drum's lectures ended with this ringing profession of faith: "Oh, I thank God with all my soul that I belong to a Church that is Indefectible, Infallible, tyrannical in its Indefectibility and Infallibility; and will never allow me to swerve an inch from the belief in Jesus, the Christ, *very man and very God, now and forever.*"

MARYGROVE
LIBRARY

CPSIA information can be obtained
at www.ICGtesting.com
Printed in the USA
LVHW020348280622
722261LV00003B/30